Management of off-highway plant and equipment

Dr David J. Edwards BSc, PhD, FFB, MCMPE, is a Lecturer and Researcher at the Off-highway Plant and Equipment Research Centre, Loughborough University, UK.

Frank C. Harris BEng, MSc, PhD, DSc, CEng, FCIOB, MICE, is Emeritus Professor of Construction Science at the University of Wolverhampton, UK.

Ronald McCaffer BSc, PhD, DSc, FREng, FICE, FCIOB, MASCE, MCIM, Eur Ing, is a Professor of Construction Management at Loughborough University, UK.

Management of off-highway plant and equipment

David J. Edwards, Frank C. Harris
and Ronald McCaffer

CRC Press
Taylor & Francis Group
Boca Raton London New York

CRC Press is an imprint of the
Taylor & Francis Group, an informa business
A SPON PRESS BOOK

First published 2003 by Spon Press

Published 2023 by CRC Press
Taylor & Francis Group
6000 Broken Sound Parkway NW, Suite 300
Boca Raton, FL 33487-2742

ISBN-13: 978-0-415-25127-3 (hbk)
ISBN-13: 978-0-415-25128-0 (pbk)

Typeset in Times New Roman by
Newgen Imaging Systems (P) Ltd, Chennai, India

Visit the Taylor & Francis Web site at
http://www.taylorandfrancis.com

and the CRC Press Web site at
http://www.crcpress.com

British Library Cataloguing in Publication Data
A catalogue record for this book is available
from the British Library

Library of Congress Cataloging in Publication Data
A catalog record for this book has been requested

Dr David Edwards was instrumental in bringing this third edition together and dedicates this book to his partner Philippa and his parents June and Tom Edwards.

D.J. Edwards (2003)

Contents

Figures

Tables

Preface

Technical improvements in off-highway plant and equipment have, throughout industry, caused a movement away from manual labour towards mechanisation. Although this has raised the potential productivity of workers, it has also necessitated relatively large capital investments in machinery and equipment, which must be operated at an economic level of utilisation if an adequate rate of return on the capital employed is to be achieved. As a consequence, an increasing number of firms are now placing far more emphasis on the selection, performance monitoring, control and maintenance of their equipment fleets. At the same time independent equipment hire businesses and DIY small tools outlets have rapidly expanded to satisfy the short-term equipment needs of businesses.

During recent years, the authors have made a special study of the management of off-highway plant and equipment and the results of their investigations have been incorporated into both undergraduate and postgraduate teaching. This has naturally directed attention towards a textbook on the subject, as it is essential that students receive guidance on the management of off-highway plant and equipment at the onset of their careers, and this book has therefore evolved to meet the needs of professionals whose roles are expanding to embrace the increasingly important management of off-highway plant and equipment.

Specifically, this third edition covers the management of equipment both within an organisation and the independent hire or rental firm, and deals with policies, strategies and organisational structures from the small to the large concern. By necessity, emphasis is placed on the financial aspects of equipment acquisition and control, and modern capital investment decision-making techniques are given special consideration. Operational management is dealt with fully, including health and safety, maintenance management, operational planning, licensing and insurance and the problems faced in overseas work. The importance of operator training and development is emphasised. Finally, the book brings together the recent developments in computer technology and its application to improving equipment management.

This book will be beneficial to students and practitioners of civil engineering, construction, agricultural, mining, forestry and building. Specifically, it will find a market among practitioners, hire and rental firms, engineers, builders, specification writers, equipment manufacturers, project managers and insurance and legal advisers. It will also serve as a useful reference for equipment managers in the industry.

David J. Edwards
Frank C. Harris
Ronald McCaffer
Loughborough and Wolverhampton Universities, 2003

Acknowledgements

The authors are particularly grateful, first and foremost to: Mr Barry Robinson, President of OPERC and Mr Richard Sharp, JCB. Our thanks are also extended to the following professional practitioners: Mr Alex Hillman, Finning; Mr Dick Stewart, Liebherr; Mr John Kerr, Scottish Plant Owners Association; Mr Bob Merchant, Contractors Mechanical Plant Engineers; Mr Michael Plummer, Marubeni Komatsu and Bomag.

We also thank fellow academics, friends and colleagues who have contributed to this, and previous editions of this textbook. In particular, Dr John Nicholas and Dr Gary Holt for their constructive comments, suggestions and checking various sections of the text. Finally, thanks to Professor Andrew Baldwin, Loughborough University, for all his support and encouragement throughout the preparation of this textbook and other research activities and to Major Martin Morris, Defence Logistics Organisation, UK Ministry of Defence for the text on Global Positioning Systems.

Chapter 1

Introduction

As markets globalise and expand, organisations face fierce domestic and international competition for complementary goods and services. In order to maintain competitiveness, organisations must look inwardly for efficiency gains. One way that these gains can be acquired is from initiatives to reduce labour costs and increase production output. The acquisition (or hire) of off-highway plant and equipment often presents itself as a viable option for organisations that strive to meet these challenges and maximise efficiency and hence, profitability. Off-highway plant and equipment includes items such as dump trucks, rough terrain forklift trucks, skid steer loaders and tracked hydraulic excavators. These versatile machines provide a considerable contribution to the industrialised economy and have been adapted for use in many bespoke operational environments, ranging from civil engineering and construction to opencast mining, forestry, aggregates and the scrap metal industry.

In demonstrating industry's reliance upon mechanisation, the chief economist at a major plant manufacturer recently estimated (*c*. 2001) that each year, over 20,000 plant items are sold to the UK construction industry alone; a market which is valued at £1.2 billion per annum. Each plant item has an average (mean) life expectancy of ten years and therefore, at any given time, over 200,000 plant items are working within the UK construction sector. As each individual item of 'mobile' equipment can cost anywhere between £10,000 and £10,000,000 (UK Sterling), it is hardly surprising therefore that capital invested in such equipment is currently valued at £6 billion (at *c*. 2000 prices sterling, when taking into account depreciation). With a further £6 billion being spent on running costs (such as consumables, labour, parts and servicing), the total invested in plant and equipment, at any given time, is therefore approximately £12 billion. Since the construction industry itself is worth £58 billion (about 10 per cent of Gross Domestic Product), the value of plant equates to almost a fifth of construction's worth.

To raise production and simultaneously lower costs, each machine must be operated and maintained efficiently and safely by competent operators. Currently, a workforce of 300,000 plant operators holds operator cards within industry, of which at least 150,000 operate within the construction sector; although, an estimated 600,000 operator cards have been issued to date. Therefore with a workforce size of 1.4 million, the ratio of plant to labour is low at 1:7 (approximately). This trend for the wider application of mechanisation and thus, more operators, is set to continue as industry strives to improve performance. Operator development is therefore an essential part of a plant and equipment management strategy.

The potentially substantial cost of capital purchase has engendered the expansion of an independent hire sector to supply the growing demand for mechanical equipment within industry. Within the UK (unlike continental Europe) plant hire companies now dominate the purchase of off-highway plant and equipment. Equipment hire companies serve two purposes.

1 They provide specialised equipment which no single company could expect to utilise fully, had the machine been purchased (e.g. harvest vehicles which are subject to seasonal demand).
2 They hold substantial stocks of everyday items of equipment, thereby relieving industry of the need to own and manage their own fleets, if they so choose.

In both instances, the organisations hiring the plant are freed from the burden of capital item acquisition which ultimately ties up organisation liquidity and prevents capital being used in more profitable areas. In addition, items are hired and off-hired as and when required therefore avoiding depreciation costs. Equipment hire firms have rapidly acquired specific management skills, which are peculiar to owning and operating equipment. This has stimulated them to operate their equipment more efficiently and perhaps more profitably than an individual company holding a small fleet.

Regardless of the plant items used or whether these are hired or owned, managers within the business enterprise are required to transform mechanical resources into products and services. By hiring the machine, the responsibility for plant management is partially transferred to a third party to maintain the item in a safe operational condition. However, the hirer can never transfer the total responsibility because ultimately the item is still used even though it is hired, hence it must still be managed. Even in the fully automated industrial environment, where manual workers are kept to a minimum level, managers will always be required. Plant managers have increasingly more autonomy, responsibility and decision-making power. Most are confronted with fundamentally important challenges on a weekly or monthly basis. On a macro level, the broadest challenge is to ensure company survival in an extremely competitive economic environment. More specific challenges include:

1 Achieving near maximum utilisation of equipment so as to generate adequate revenue and a sufficient return on capital.
2 Delivering a high quality service to clients in order to secure future contracts.
3 Managing cultural and technological changes efficiently and effectively.

Consequently, managers who hold large (capital intensive) equipment fleets must give very careful consideration not only to initial equipment selection but also the method of acquisition, the monitoring of usage and performance, training of plant operators and the maintenance of their fleet. A holistic range of skills and knowledge is required by the competent plant manager.

Managers (not necessarily the plant itself) are the most valuable resource in most businesses since they share the responsibility for off-highway plant and thus company performance. However, the managerial resource can quickly depreciate in the absence of sufficient and constant nurturing. It takes years of dedicated work to finely tune a manager's ability yet such can be depleted in a relatively short time span particularly if the manager seeks alternative employment or fails to keep abreast of the latest technology. An exact cost of enhancing the manager's skills is difficult to estimate and largely depends upon the nature and scale of the business. Nevertheless, the cost of under investment is considerable and will mean the difference between a successful and failed business venture.

About this book

This book provides a comprehensive treatment of off-highway plant and equipment management. The body of knowledge described as equipment management comprises management organisation, economic evaluations, budgetary control and costing, cash flow, operations management and financial management. It also involves maintenance and the control of maintenance costs, the use of information technology, human resource management and knowledge of health and safety, road transport laws and insurance. The book does not aim to provide a detailed description of individual plant items, equipment operation and associated machine performance. Manufacturers provide the finest source of such literature particularly since new machines and innovative technology are constantly being developed.

The book does provide readers with a concise, but thorough, guide to plant management, presented in an interesting fashion. Throughout the text an applied treatment of plant management principles is given in a straightforward manner. A central feature of the book is its multidisciplinary approach. The book is suitable for both introductory and advanced courses at an undergraduate, professional and postgraduate level in construction, civil engineering, agriculture and mining management. However, it will also be relevant to practitioners and students on full- or part-time courses and distance learning packages in engineering, operations management or any other course that contains an element of business and management science.

This book consists of four distinct sections, namely organisation for business; investment, procurement and business management; operational management; and financial and budgetary control. Each area is now briefly introduced in turn.

Part I: Organisation for business

This section considers the options for equipment acquisition and supply and looks at the organisational structure and business strategy needed to successfully manage the plant holding. It covers external equipment hire and rental firms and the internal plant and equipment department of an organisation operating in profit- or service-centre markets. Finally, the section provides an overview of the most popular equipment types available.

Part II: Investment, procurement and business management

This section deals with the economic criteria for evaluating investments in equipment, including the effects of corporation tax, capital allowances and inflation. These economic analyses are also applied to the various forms of ownership. A detailed procedure for selecting the most suitable equipment for acquisition from a range of available alternatives is described, and the calculation of an economic hire rate taking account of ownership costs is given in a detailed example.

Part III: Operational management

This section on operational management considers the more practical daily roles and responsibilities of the plant manager. Legal and contractual insurance requirements, especially hired plant and equipment, are dealt with, and licenses for plant operation on the public roads are reviewed. The various strategies available for effective control of equipment maintenance and its costs are also discussed. Health and safety requirements and plant operator training and competency issues are summarised. The section concludes with guidance on operational planning and international operational logistics.

Part IV: Financial and budgetary control

This section deals with budgetary control and costing, cash flow and financial management as applied to hire companies and equipment divisions. The use of information technology in equipment management is also explained.

Part I

Organisation for business

Chapter 2

Organisation and management of equipment services

Introduction

Equipment items are typically acquired gradually by a commercial organisation to support its business activities, until the holding eventually reaches a size where it becomes necessary to organise the fleet into a separate unit. This marks the beginning of a plant division or even subsidiary, which may continue to develop, supplying the needs of the parent company. In some cases it may become a fully independent concern.

Today equipment ownership is not a fundamental requirement, as a vast selection of machinery is available for hire as an alternative; nevertheless, many enterprises choose to purchase some of their needs, for reasons of convenience and prestige. Such a decision should be considered carefully since plant ownership is a capital-intensive business and requires a relatively large central organisation to provide all the facilities for maintenance, cost accounting, hiring, etc. Indeed the following questions should be answered before any item of plant is acquired, namely:

1 Is ownership of that item of equipment fundamental to the operations of the business?
2 Will the capital locked up in the equipment generate an adequate rate of return compared with other forms of investment?
3 Is purchasing the equipment for direct ownership the only profitable way of obtaining and operating it?

Unless the reply to these questions is unequivocally positive, some other sound commercial reason should be established before authorisation to acquire the equipment is granted.

Clearly it is a major management task to lay down principal policies for the supply and organisation of equipment holdings, so that the objectives for the whole enterprise may be achieved. The consequence of these policies will stand as a record for later revision as required. If this is not done, the plant fleet may become a hidden and ever-increasing drain on financial resources. Furthermore, the strategy for achieving profitability for commercial equipment operations may not always coincide with that of other parts of the business.

Equipment acquisition policy

The means of obtaining equipment may be broadly classified as follows (refer to Figure 2.1):

1 own all equipment – including hire purchase, straight purchase and leasing;
2 hire in all equipment;
3 combine own and hire.

Each method will make special demands on the use of capital and resources.

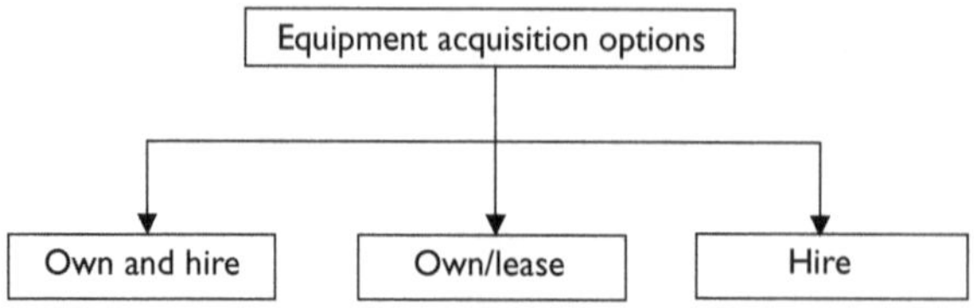

Figure 2.1 Equipment acquisition options.

Own all equipment

The policy practised by many enterprises is to purchase, or lease long-term, most of the equipment needs and thereby provide availability at all times, with the added advantage of the prestige attached to demonstrating the use of owned equipment. However, much capital will be locked up in the plant, which must become capable of generating a sufficient rate of return. A major disadvantage of this strategy is the problem of maintaining adequate levels of utilisation. Equipment holdings are usually built up to service a growing demand, and will become a heavy liability in the case of an economic recession. Any available work may then subsequently need to be undertaken to sustain the fleet, since equipment cannot easily be sold in a declining market.

Hire all equipment

Many specialist hire/rental firms offer the supply of plant and equipment on the open market. To take advantage of this facility avoids both the responsibility for maintenance and the tying up of capital. The equipment may be rented for a specified period and hire charges minimised by standing off-hire all unwanted items. In many instances the plant operator is also provided by the equipment supplier.

The main disadvantage of hiring is that hire rates depend on market forces and suppliers are largely beyond the control of the hirer, except for limited negotiation between competing firms.

Combination of own and hire

A mixed policy of owning and hiring plant may be the preferred option. For example, regularly required items might be purchased and hiring adopted only to smooth out demand.

Establishing the objectives of the enterprise

The equipment service/hire business, like any other, survives by combining skills and talents into an organisation that can produce services in sufficient quantity to satisfy the material desires of the community in which it exists, and at the same time provide a sufficient return on the capital invested. In practice many other objectives need fulfilment including satisfying the desires of the shareholders, management and workforce to see the enterprise grow and continually increase turnover, to become well known and respected in its field, to operate in more stable markets and to keep up with competitors: these are but a few aims. The ability to achieve such objectives is continually influenced by business changes as indicated by pressure on profits, supply outpacing demand or vice versa, competitors being more successful and through alterations in customer tastes and requirements.

In order to operate within such a market framework the enterprise, or more correctly its senior management, must establish clearly:

(1) *The appropriate kind of business structure.* For example, (a) only provide internal plant services at rates of hire reasonably comparable with those available in the open market; or (b) organise plant holdings as a separate entity responsible for generating its own capital and profits, with the freedom to hire equipment both internally and to the outside market.

(2) *The type of services to be offered.* For the provision of plant as a 'service' arrangement, management must decide how much capital should be invested in equipment to service its own internal needs, thereby setting indirectly the limits on the proportion that will be self-owned. The 'profit-centre' strategy requires plant to be managed as an independent organisation, with a specific share of the market for equipment hire established as a major goal. Critically for the latter, this involves taking decisions on whether to concentrate on local, regional, national or international markets; construction, mining/quarrying, agriculture, transport or industrial sector needs; earthmoving, craneage, mobile, small general plant or specialist equipment, DIY equipment; commercial maintenance services, etc.; as well as the corresponding location and organisation of stockyards, rental outlets and service facilities. A strategy for innovation, quality, delivering on time and developing good and continuing relations with clients also has to be properly formulated.

(3) *The desired share of the market.* The potential market share needs to be based on a careful evaluation of economic trends and opportunities, married to a detailed assessment of the enterprise's capabilities and likely potential for assembling the appropriate commercial, managerial and technical skills, resources, etc.

(4) *The appropriate skills and knowledge base required.* Know-how in people is vital in creating sales which increasingly rely on providing added value, developing efficient and innovative equipment and services, effective processes and management systems.

(5) *The possible changes in and fluctuations of the market in future years.* Political and economic shifts in consumer outlook, particularly concerning the environment and quality, vary both between different markets and over time. Fluctuations will affect the demand for plant and equipment, and management should always be looking ahead at the potential consequences. There is likely to be more opportunity to adjust for a commercially independent organisation which can set its own objectives and decide on strategies, but even a small service plant department should maintain commercial viability. Whether the plant holdings are organised into a 'profit' or 'service' centre, they must be profitable.

By setting these major objectives, management is in a better position to structure the organisation to meet the challenges.

Management structure

The appropriate management structure for a plant and equipment organisation will depend upon the nature and size of the firm's business activities. In particular, an independent rental or hire firm will require all the management functions of a market-oriented company, as shown in Figure 2.2. An internal plant division merely providing a 'service', will generally be integrated into the parent company's activities, and functions such as purchasing and financial accounting may be outside its responsibility.

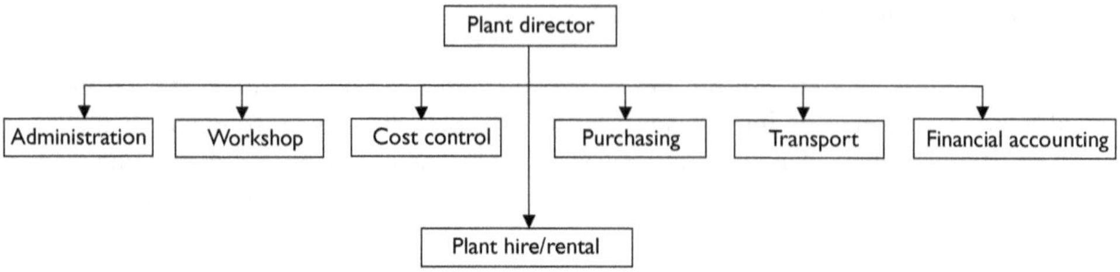

Figure 2.2 Management structure of a plant hire business.

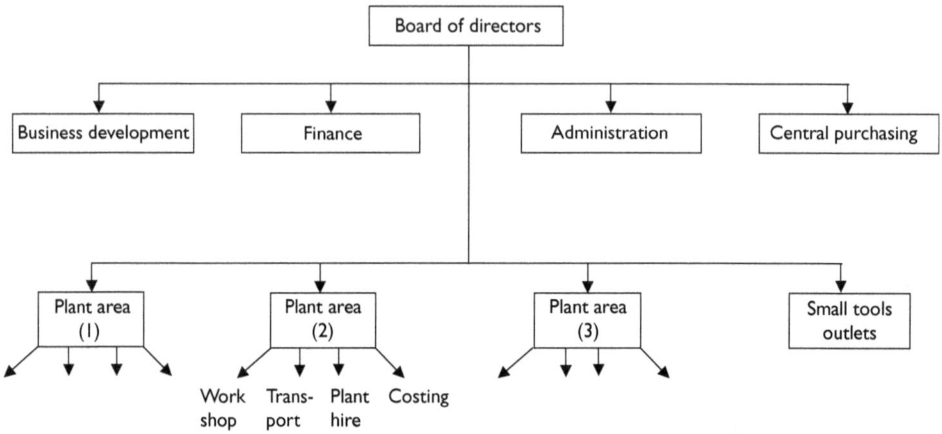

Figure 2.3 Management structure of a regionalised plant holding.

For both types of business, the need to decentralise into geographical regions, or even major equipment categories, is a further complication. Firms tend to make individual depots or sites responsible for their own business activities when faced with this situation, although responsibility for overall company policy, business development, major purchasing and financial accounting may remain at head office (see Figure 2.3).

Typical management functions and departments

The managing director

The managing director, as chief executive officer, answers to the owners/shareholders through a board of directors which sets the objectives of the business and assures that necessary strategies are adopted to ensure that the enterprise will survive and prosper. The managerial functions involved in all but the smallest firms require that much of the day-to-day responsibility for running the various departments is delegated, although ultimate responsibility for the success or failure of the business must lie with the managing director and the board. For example, important matters of the company policy – expansion plans, development of markets, sourcing of capital, capital expenditure, and purchase and disposal of equipment items – are usually decisions led by the managing director, whose responsibilities may also include liaison with key customers and setting cost and financial budgets for each department.

Board of directors

The bigger company is usually more decentralised and is managed through the board of directors. The board is responsible to the shareholders for corporate management, strategy and governance, with the managing director holding overall responsibility for policy execution and usually chairing the company's business. The board will commonly have a varied membership with separate executive directors appointed to direct each of the major functions, such as business development, administration, technical services, accounts and workshop operations, depending on how the board decides to allocate responsibilities. Such directors may or may not be shareholders according to their financial ability to assemble a major portfolio of shares.

A very small minority of organisations, through acquisition activity, have been able to develop a complex group holding of companies with the main board of directors comprising executives empowered as managing directors to oversee the performance of a subsidiary or major component of the company. Sometimes non-executive directors are invited to serve on the main board offering specific expertise emanating from influence in banking, politics, the legal profession or business development, with others perhaps nominated by major shareholders to look after their interests, audits etc. Usually, a chief executive officer is appointed to steer the group and chair the main board to the satisfaction of the owners, that is, shareholders, but sometimes a separate chairperson is nominated to preside over business affairs thereby containing excessive independent action of the chief executive.

Administration

The administrative function will grow with the size of the business and in large firms is subdivided into separate elements. The list of duties includes health and welfare of personnel, employee training, provision of social facilities, a postal service, legal and insurance advice, negotiation of wages and salaries, conditions of employment, personnel record keeping and maintenance of the physical assets. Responsibilities commonly extend to research and development, information technology management, quality management and audits, and safety management, although some of these are becoming sufficiently important to warrant separate control under a services function. Authority is vested in a personnel/administration manager with subordinate managers responsible for the different elements.

Financial accounting

The company accountant is responsible for the payment of invoices, receipts from hire sales, control of cash and bank overdrafts, and preparation of the trading, profit and loss accounts and balance sheet in accordance with the Companies Acts. The accountant has an important function and often works closely with the managing director in controlling the overall financial affairs of the enterprise. For example, in making decisions on the type and source of capital for major purchases, assessing the financial viability of an expansion programme and preparing the company financial budget for the year ahead.

Purchasing

The buying function is responsible for obtaining quotations for materials, supplies and other consumables used at the depot and on plant and equipment located in the field but maintained and supplied from the central or regional depot. Thus, advantages of centralised purchasing may be realised from:

1. The ability to obtain discounts from bulk purchasing.
2. The efficiency generated by adopting standard procedures.
3. The ability to monitor the quality of supplies.
4. The experience gained from evaluating the commercial operations of suppliers and subsequent development of preferred suppliers in a committed network.
5. Centralised administration facilities.

However, when the plant need is relatively small with items maintained on site, a separate purchasing department is hardly necessary and consumables are simply charged to the workshop maintenance account.

A secondary purchasing function may be participating in the purchase and sale of equipment, especially inviting quotations and assessing the commercial details of a transaction, although the advice of other departments will also be involved at this level of plant procurement and disposal.

Cost accounting

The cost accountant collects and interprets data from other parts of the business and prepares targets, in the form of budgets, against which costs and revenues may be monitored. This information is used by departmental managers for control and updating purposes. In particular, all the data required to prepare hire rates for equipment must be recorded and returns are therefore required from all departments, including the hours operated weekly for each item.

Workshop control and maintenance

The workshop management function primarily delivers maintenance and servicing of equipment. This includes provision of a store of consumable materials and spare parts, with accompanying appropriate stock control procedures, although the actual purchasing responsibility usually lies with buying. Costs incurred include the wages of fitters, mechanics, electricians and other operatives needed to perform servicing and repair duties, plus the costs of tools, materials, mobile workshops, general overheads, salaries paid to staff and foreman, and equipment inspection and administration. The budget is allocated from past records of the equipment holdings and any additions that can be foreseen. Hence maintenance records of each item must be rigorously updated to facilitate monitoring of costs against the budget. As a rough guide, maintenance facilities for about 5 per cent of the fleet should be provided.

The workshop manager's function may be augmented by a field specialist to provide advice to sites on the use, operation and routine maintenance of plant and equipment. Indeed the duties may be complementary, where a comprehensive system of mobile maintenance to sites is supplied. A complaints service may also be provided and the role then needs to be carefully integrated and co-ordinated with the business development function.

Transport

Most equipment is unsuitable for travel on public roads and must be transported from site to site on trucks and lorries, co-ordinated by a transport supervisor through the workshop and hire desk. The costs allocated include the running costs of the transport fleet, such as fuel, maintenance, servicing, drivers' wages, supervisors and administration staff salaries, the capital cost of the transport fleet and overheads. The responsibility for servicing and maintaining the transport fleet will generally remain with the transport manager, but execution of the work may be undertaken by the workshop and subsequently charged to the transport account.

Hire desk: business development

The business development function is primarily concerned with marketing, (particularly developing growth by identifying and seeking out new opportunities), promoting the firm's products and services, including responsibility for technical appraisals and supply of advice to customers on all aspects of the firm's business from procurement to final delivery. A well-founded department would incorporate a wide variety of the company's expertise acting coherently in a business development role, able to offer clients and customers scope/feasibility, operational, legal and contractual services and consultancy, etc. The conventional 'hire desk' provides some of this function, principally the limited aspect of selling.

The asset register

To assist co-ordination across the business, most key data on each equipment item are recorded in an asset register, either manually or more advantageously as stored data in computer files, typified by Table 2.1. The basic information required should include for each item, a code number, a registration number, make, model and short description. Additional data and separate reports should be added as necessary. For example, the financial accountant requires data on purchase date and price, planned life, depreciation method, book value, depreciation charge and depreciation to date whilst the hire desk requires current utilisation factors, hire rates, budgeted earnings, actual earnings and actual costs. An inflation index for the particular equipment category is also useful in setting new hire rates in times of inflation. The workshop needs current data on location, base depot, planned operating hours and actual operating hours so that effective maintenance can be monitored.

Furthermore, an up-to-date asset register assists in programming safety inspections, required for insurance and legal purposes. Last but not the least, the managing director must be constantly aware of an item's profitability and utilisation level.

Size and distribution of firms

A considerable variety of equipment is available, as self-owned or for hire, in the different industrial sectors. Table 2.2 illustrates data for the 3,000 firms or so in the construction rental sector and may be typical. It indicates that large organisations with 60–300+ employees, although representing only 5 per cent of total firms account for slightly more than 50 per cent of the total value of business. The main firms are represented by the Construction Plant Hire Association (CPA) and Hire Association of Europe (HAE), these being the member organisations offering contract conditions guidance, economic forecasts and general commercial advice and assistance.

Table 2.1 Example of an asset register

Machine number	*Machine description*	*Purchase date*	*Purchase price (£)*	*Scrap/resale value (£)*	*Machine life*	*Type*	*Depreciation charge (£)*	*To date (£)*	*Written down value (£)*	*Machine hire rate (£)*	*Budgeted earnings to date (£)*	*Actual earnings to date. this month (£)*	*Total cost to date (£)*	*% P/L on earnings*	*% Utilisation*		*Location*
															Year	*Month*	
601	Terex TS14/70	09/1984	100,000	20,000	5	DBAL	16,000	60,000	4,000	45.36	4,600	5,000	4,500	10.0	82	85	Bristol
602	CAT 633C Scraper	06/1984	120,000	20,000	5	DBAL	20,000	55,000	45,000	47.49	4,000	3,950	3,500	11.4	75	76	Birmingham
603	Terex TS14/75	11/1985	118,000	18,000	5	DBAL	20,000	40,000	60,000	48.33	5,000	6,000	5,900	1.7	76	73	Wolverhampton
604	Terex IS14	11/1985	118,000	18,000	5	DBAL	20,000	40,000	60,000	48.33	4,900	4,000	4,100	−2.5	70	67	Loughborough

Table 2.2 Plant and equipment businesses in the construction sector

Percentage of total firms	*Number of employees*	*Percentage of total turnover*
60	1–10	10
35	10–60	40
5	60–300+	50

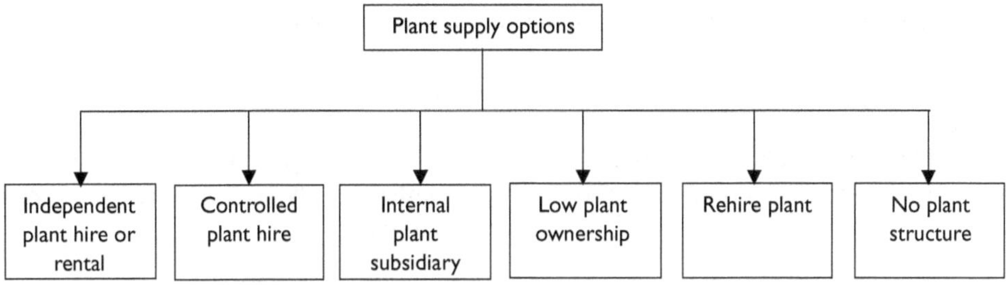

Figure 2.4 Plant supply options.

Alternative options for equipment supply

Within the above market framework there are many enterprises that have plant holdings, some quite large, which are not subject to open competition, and it is important that the capital invested be used efficiently. Notably, where the proportion tied up in equipment is high, it is obligatory to try to maximise the profit on the investment; otherwise the capital would be better used elsewhere in the business. Any other approach will carry serious dangers. For example, if equipment is purchased to provide a service to other parts of the business and charged below market rates, the resulting low bids must lead ultimately to the other parts having to earn excessive profits to generate an acceptable rate of return on the total capital employed.

Recent trends suggest that plant managed as a 'profit centre' is likely to perform better than that organised as a 'service centre' as far as profitability is concerned. In the case of the former, the rigours of the market place ensure that only equipment which can show a high level of utilisation and/or profit throughout its working life is purchased. In addition, the constraints are such that the costs of maintenance are controlled so that neither too much nor too little is undertaken. For the 'service centre' however, operating costs are ultimately met by the enterprise as a whole, and there is a tendency both for maintenance facilities to expand and items to be purchased, with little regard to levels of utilisation.

Business arrangements have evolved to accommodate different commercial needs (refer to Figure 2.4), each with consequences which affect the efficient use of resources.

Equipment organisation in practice

The typical distribution of plant and equipment holdings is detailed in Table 2.3. Almost half of these organisations operate their plant holdings as a profit centre (I and II). A minority

Table 2.3 Typical equipment holdings

Type of plant organisation	*Per cent of firms*
(I) Independent plant hire	25
(II) Controlled plant hire	20
(III) Internal plant subsidiary	10
(IV) Rehire organisation	15
(V) Low plant ownership	10
(VI) No plant structure	20

employs a service system of internal plant hire (III), these being large general contractors with sufficient work in their own organisation to keep utilisation at levels which are economic and profitable for the firm. The rehire and low or no plant options (IV–VI) are clearly favoured, especially for small businesses or those which predominantly rely on the plant hire segment.

In addition to the above organisations, the DIY small tools and equipment sector is today a major market, with several thousand shops and outlets supplying a wide range of small items, from garden tools to small-scale industrial equipment.

Independent hire or rental shop

This type of organisation covers both the independent plant hire business, including DIY small tools outlets, and the plant hire subsidiary operating under the umbrella of a holding company. Equipment is supplied to the market to provide a satisfactory rate of return on the employed capital. Decisions on equipment purchases and plant holding policy will be taken by senior management. Normally, the subsidiary will be given a name which is not associated with the parent firm, so that other similar companies will not be discouraged by advertising a competitor's name on site.

Plant for hire or rent will be purchased to make a profit based on its utilisation potential and the maintenance record experienced with similar items, together with an assessment of the hire rates likely to prevail in the future.

Controlled plant

The plant fleet may ultimately become so extensive that, to maintain effective control, the holdings are incorporated into a subsidiary division. The first priority may be to serve the equipment needs of the parent business at a profit, but in order to maintain high levels of plant utilisation and thereby maximise profits, items may be hired out to other users. As a rule of thumb, the ratio between 'hiring internally to the parent organisation' and 'hiring externally to the market' should not be less than 2 : 1.

However, the tendency with this system is for the equipment to be hired out to the market when the rates are attractive, since there will be a demand for such services and the required utilisation levels can be more easily achieved this way. Consequently, there is a danger that the internal needs may be neglected, and items will not be available at the right time for the organisation's own uses. Furthermore, the servicing and maintenance requirements of equipment hired internally may have to take second place to that required on the open market.

Internal plant subsidiary

The dilemma of servicing two different types of client presented by a controlled plant policy is eliminated when the activities of a plant subsidiary are restricted to internal hire only. This system often results in plant hire rates which bear little relationship to market rates, as the type of equipment items and utilisation levels are dictated by internal demands. Nominally, the plant subsidiary is required to achieve a set rate of return on the capital employed. Sometimes, however, the targets cannot be achieved and the deficits must be covered by the parent organisation. Such an arrangement may produce a management team not held in the same regard as profit-oriented parts of the enterprise, with a consequent loss of influence and confidence of the plant manager. Decisions on plant purchases and control may then increasingly be imposed by the operational side of the business, to compound the difficulties of making profits from the equipment holdings.

Low plant ownership

Some businesses keep very small equipment holdings, on the grounds that achieving profitability from plant ownership is relatively less rewarding than other commercial activities. A small depot may be maintained to provide small items only and most of the problems of owning equipment are avoided. This system, of course, relies heavily on the facilities provided by the plant hire sector. The availability of specialist equipment items can influence the work load and contract type open to the enterprise, and so could affect the success of this sort of policy.

Rehiring plant

In order to reduce the administrative duplication of each site (or contract), obtaining and then invoicing for payment the plant requirements, some enterprises operate on a basis similar to the low plant ownership arrangement, but provide a centralised service of hiring-in all equipment and passing it on to the sites. The main advantage lies in the ability of a centralised administration to negotiate favourable terms and discounts with regular hire firms. Besides economy, some co-ordination of plant hire requirements across the company can also be achieved and so accommodate the transfer of equipment items from one location to another. An extension of this system is the recent emergence of rehiring businesses supplying equipment to the hire sector itself.

No plant structure

Management sometimes takes the view that an unstructured organisation with respect to plant holding will serve needs best. Several arrangements are possible. For example, individual contracts may purchase equipment and be credited subsequently with nominal resale values when the plant leaves the site. In this case care has to be exercised in assessing equitable sums when purchases and resales are internal transactions. This method is usually confined to special items, such as grouting pumps, cableways, etc., which are usually sold off when a contract has no further use for them. In conjunction with this system, more general items may be moved from site to site without a formal charging procedure. Plant is costed as an overhead to the contract on an arbitrary basis, but these policies clearly carry the risk of not forcing the plant to make a sound financial contribution to the company's activities.

Chapter 3

Marketing and business development

Market planning: an introduction

The principal purpose of a plant organisation is to supply its equipment services to clients at a profit. However, the market for plant varies dependent upon both the opportunities for specialisation and the quality of service demanded. The organisation structure may vary from the rental company operating in the open market to the 'service' plant division typically found in a contractor or public sector department, and not least, to the many owner–driver businesses. It is important that management properly define the business aims and objectives, so that a suitable organisation may be assembled to operate in the appropriate market segment(s). In this respect development of the business involves establishing the customer's plant requirements, so that the enterprise can organise itself in the best possible way to satisfy those demands, commonly referred to as marketing.

Marketing is defined by the Institute of Marketing as:

> The management function which organises and directs all those business activities involved in assessing and converting customer purchasing power into effective demand for a specific product or service, and in moving the project or service to the final customer or user so as to achieve the profit target or other objectives set by the company.

Importantly the marketing function extends beyond the actual selling function to fully embrace the process of identifying and seeking out new opportunities and promoting services, as well as taking responsibility for technical appraisals and supplying advice to customers.

The market and business development strategy

To be confident in establishing realistic market objectives, senior management must be prepared to undertake the most thorough investigation of the business organisation and opportunities, broadly involving two separate stages.

1 formulating a market forecast;
2 assessing the strengths and weaknesses of the organisation.

In conjunction, these two exercises will facilitate the setting of new policies (see Figure 3.1).

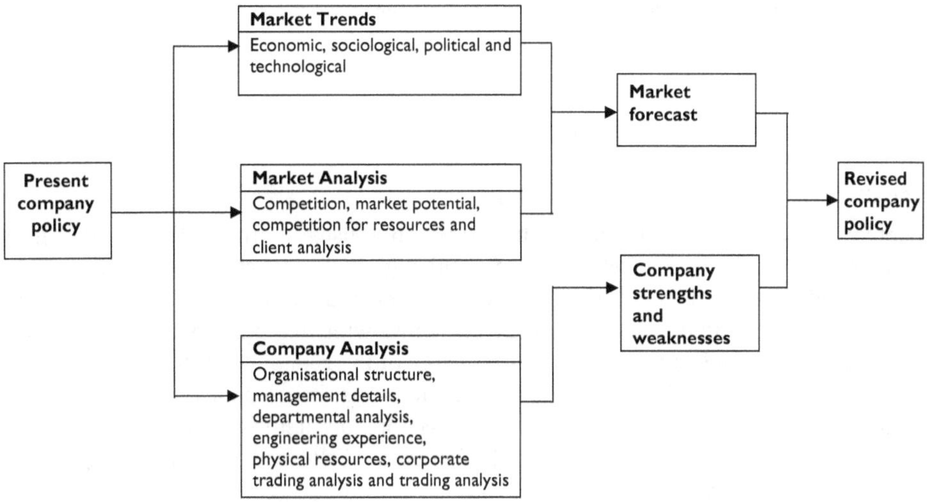

Figure 3.1 The market forecast.

The market forecast

The market forecast should endeavour to seek out the wants and needs of the market for equipment services and should be a systematic and continuous process, executed religiously, if the firm is to survive and prosper. The task can be broken down into separate stages from which information is finally synthesised to produce the new policies and strategies. The typical main areas are shown below.

Analysing the competition for services provision

A brief survey of other equipment hire and rental service suppliers may reveal segments of the market which have not been fully exploited or, conversely, should be avoided because of fierce competition. The main points to determine are the strong and weak areas of the hire market, including the following aspects:

1 The enterprises's market share of the different lines of plant and equipment held, together with an analysis of the recent performance of each type with respect to growth and profitability.
2 The market share, growth in turnover and profitability of the major competitors, noting the areas of interest for each.
3 The margin of difference between internal hire rates and those of the competition, to give a guide to the improvements required.

Analysing the potential market for equipment services

Clearly, the demand for the services of plant and equipment will be reflected in the general level of business activity. The aggressive enterprise will research the major areas of the national economy (and international opportunities where appropriate) to seek out those sectors of potential growth. This should include analysis of the regions or projects designated by government for special development and the private sector developers investing heavily in new facilities.

Analysing the competition for resources

FINANCE

Equipment requires heavy capital investment, which is usually principally provided from internal private resources and retained profit, although hire purchase, leasing, bank loans, etc. also need to be considered. Availability of loan capital in particular is likely to fluctuate according to the fortunes of the national economy, with unpredictable changes in the interest rate, when lenders may prefer other sectors of the economy. Furthermore, a sound financial record of profitability, with mortgageable assets, is essential for favourable consideration by the banking and financing sector. Before any item is acquired, it should be remembered that a machine once purchased often cannot be turned quickly into liquid cash assets to deal with a crisis.

PLANT, PERSONNEL AND PREMISES

Few enterprises possess the resources or expertise to operate in all sectors of the plant and equipment services markets. Therefore, external factors which would affect the ability to compete must be defined. For example, the reliability of the various manufacturers and supply agents should be assessed, since the quality of back-up services and availability of spares will have a considerable bearing on competitive performance.

Plant will usually last longer and be less costly to maintain if the machine operators and servicing staff are well-trained and responsible. There is always competition for such skilled personnel, and a business not prepared to train, educate and properly reward the workforce should avoid sophisticated and technically complex equipment holdings.

Extra premises may be required for new developments such as a rental shop or area division, where often the depot must be located near the main market, such as a large town. However, sites which can provide room for expansion, good access and security will be in demand from other firms and industries. These aspects are often overlooked when expansion programmes are put into operation.

Analysing the client

Some clients are better 'payers' than others, so helping reduce the need for cash or overdrafts. It is essential to be cautious, even though there may be an apparently lucrative market for certain equipment lines. Clients may be so slow and awkward in their attitude to payment that their avoidance would be preferable.

Strengths and weaknesses of the organisation

It is essential to consider the ability of the organisation to cope with a change of direction or expansion. Its structure must be examined for both the effectiveness and strength of each element.

The corporate analysis

ORGANISATION STRUCTURE

Most enterprises have a definable family tree which represents the official structure of the management organisation. In practice, the actual lines of command and communication are likely to be more subtle than those formally recognised. However, this family tree is a good starting point in highlighting potentially weak structural arrangements.

MANAGEMENT DETAILS

The quality of present managers will be tested when entering new markets. Much information is often held in personnel records on such matters as salary, qualifications, education, training and experience. These data help to identify potentially strong management areas and those which have failed to develop a healthy ladder of achievement on which the younger employees can gain experience. If the process is repeated for each operational element, gaps and stagnant areas become apparent.

FINANCIAL AND OPERATIONAL CONTROL

This review should be extensive and probably not undertaken until policies have been made tentatively. The most likely candidates for investigation are the accounts, administration and equipment servicing functions, since they tend to be labour-intensive, and hence would have difficulty in responding quickly to changes in direction. Such functions are significant contributors to overheads, which may need to increase if the enterprise expands into new markets. Overheads should also be borne in mind when moving from a fairly low technical market – say small machines – into a specialised market requiring high technical competence and support.

ENGINEERING EXPERIENCE

Management and operational control surveys will yield useful information about the nature of the business and its employees. Since plant services supply requires that good managers should also be good engineers, any change in policy should spring from a sound base of experience: it is far too risky to rely entirely on imported skills when undergoing change. Therefore, a careful analysis is required of the existing skills to see whether they will provide an adequate basis on which to build. In particular, staff expertise is likely to be severely tested when policy changes involve the introduction of new equipment lines or when the firm decides to decentralise and establish depots sited away from headquarters.

PHYSICAL RESOURCES

Putting new objectives into practice may necessitate new depots, outlets and storage facilities. However, the acquisition of land and the construction of new facilities take time, are expensive and demand careful planning of the location. In addition, new and different equipment items may require new maintenance facilities, which may be costly and beyond the knowledge and experience of the management and workforce.

CORPORATE TRADING ANALYSIS

The following financial ratios yield important information in assessing the financial strength of the business and comparing its performance with that of major competitors:

- return of capital employed;
- profit on turnover;
- turnover of capital;
- growth in capital employed and in net profits;
- current assets to current liabilities;
- stock values to sales
- debtors to sales
- profit per employee

(stock values to sales, debtors to sales, profit per employee: converted to time periods.)

By comparing figures over the past five years with other organisations in similar fields, some judgement is possible on the viability of the enterprise and its ability to take on new ventures successfully.

TRADING ANALYSIS

The trading analysis means looking at individual equipment lines in a fair degree of detail. The types of questions to be asked are:

1 What trends in profitability – say during the past five years – can be seen for the various equipment types and lines?
2 How did actual profit compare with estimates?
3 How has inflation affected costs and what was the policy towards hire revenues?
4 What effect would changing the mark-up included in plant hire and equipment rental rates have had on turnover and overall profits?
5 How did maintenance costs, actual machine life and utilisation levels compare with estimates?

Trends affecting the forecast

The stage of proposing any changes in company policy, as shown in Figure 3.1, has now been reached. Once the facts are known, experienced senior managers will usually see what changes need to be made. When these tentative proposals have been put forward, it should be understood that new facts will emerge and errors in the forecasts will appear. These are inevitably caused by political, environmental legislative and sociological changes, shifts in the market outlook, technological developments and economic influences. The effects of these movements are difficult to quantify, but should be kept under cautious review, the policies being adjusted where necessary. However, care should be taken not to overreact to new events, as this can cause loss of confidence at middle management level.

Implementing the market plan

The business development function

All the major technical and managerial elements required to draw clients and customers into meaningful contracts, that is, those traditionally associated with the 'hire desk', need augmentation to include scope/feasibility advice, operations planning services, buying, legal and contractual matters, etc. For best practice, co-ordination would be directed under the umbrella of the business development function, where a wide variety of the enterprise's expertise could be made available to potential customers, thereby more effectively helping to engender good business relations and foster repeat work.

Promoting the services and satisfying the customer

Once the plant and equipment lines and the desired market to service have been established, it becomes vitally important to increase the awareness of the potential customer. This may be achieved by a variety of advertising methods coupled with fostering good public relations. The latter will probably only bring results in the medium-to-long term and should include providing clean, reliable and well-maintained equipment with an efficient back-up service of spare parts and advice. Co-operation with the client or customer is always helpful as contracts often involve slight delays and changes to the original requests for plant operation and hire. These are not always detrimental to the operating costs of the firm, and the goodwill generated will help in the future.

Many customers and clients are impressed by 'added-values', and clearly the firm with back-up services of experienced and well-qualified staff in servicing, maintenance, law, insurance, technical advice, etc., will be a more credible company than one without such facilities.

Chapter 4

Popular equipment types

Introduction

The range of off-highway plant and equipment presently available to the plant manager, in terms of make, model and size, is considerable. This diversity can in part be attributed to industry's insatiable need for mechanisation. Over time, mechanical engineering design has developed new innovative items to meet this increasing market demand. Originally, plant engineering design concepts (*c*. 1900–1960) were largely mono functional, that is, each plant item had its own 'distinct' application; either in the agricultural or major earth moving sector of industry. Modern engineering design concepts (*c*. 1960 onwards) have taken a more holistic view to engineering requirements, and have subsequently pioneered multi-functional 'utility' machines. That is, machines have evolved to be able to perform a variety of tasks in various industrial applications. For example, the wheeled backhoe loader can be fitted with a wide range of machine attachments (impact hammers, road sweeping brushes and so forth) which enable it to be employed in different working environments.

To those who are not experienced in plant and equipment management, choosing the equipment type for a given application can be a daunting task. There are many aspects to consider, such as machine cost, productivity, adaptability and efficiency, all of which ultimately influence business competitiveness and individual project profitability. Choosing the wrong plant item can therefore be both inconvenient and costly. Selecting the correct equipment for a given situation requires a combination of general knowledge (of the equipment) and more specific, up-to-date information on, for example, machine configurations and performance. The latter is difficult to achieve from any single reference source, since new machines and models become available on an almost continual basis. However, such information can be found relatively easily by contacting plant manufacturers directly and requesting machine performance specification literature. This is also the sensible approach because, in theory at least, one will obtain the latest technical or performance data. Trade articles are also an important supplement to manufacturer literature since comparisons between alternative machines' performance tend to be a regular feature.

This chapter presents an introduction to some of the more popular off-highway plant items available. The aim of this introduction is to highlight the range and diversity of plant available. A more detailed guidance on individual machines may be obtained in *Modern Construction and Ground Engineering Equipment* (Harris, 1994).

Bulldozers and tracked loaders

The first 'bulldozer' was called a 'crawler tractor' and was manufactured in 1904, resulting from the efforts of an American engineer named Benjamin Holt. Holt later bought-out his

Figure 4.1 Bulldozer.
Source: This figure is reproduced with kind permission from Finning (Caterpillar) UK Ltd.

rival Daniel Best and the company Caterpillar was born. Early machines were mainly used for agricultural purposes and in a physical sense tended to resemble steamrollers. Whilst the bulldozer was originally developed for the agricultural industry, today's machine is firmly embedded within the earthmoving industry (more commonly known as 'muck shifting') and is designed to clear ground using a large vertical, curved blade situated at the front of the machine. The modern machine is generally restricted in its application to roadway/motorway grading, trenching and maintenance of haulage roads but other industrial tasks include landfill, demolition, site clearance and logging (Figure 4.1). Bulldozers can also be used to tow other vehicles such as scrapers, lorries and smaller plant items such as towed sheep's foot compactor rollers. The number of companies that manufacture bulldozers is limited due to the fact that the market is small and the competition fierce, but the predominant manufacturers include Komatsu, Caterpillar, Case, Liebherr and Fiat Hitachi.

Although both wheeled and tracked versions of the bulldozer are available, the tracked type dozer is more common. The philosophy of the tracked plant item is that its tracks run continuously along the ground, in segments, driven by high-level sprocket wheels. The benefits of this are increased traction and, as a result of maximising the area of track in contact with the ground, minimal force resulting from machine weight. That is why tracked machines, despite their weight, can work in very poor ground conditions. A hybrid variant to standard tracks is 'swamp tracks' which are extra wide tracks that allow the machine to work on water-logged ground.

A range of blades and other attachments such as a rear-mounted ripper (for removing stubborn objects or to rip rock into small pieces) can be fitted to bulldozers. Other features may include the optional use of a laser mast so that the machine's doze level can be precisely controlled.

When purchasing (or hiring) a bulldozer, several key performance characteristics should be assessed and compared in order to select the most suitable machine for the given task. These characteristics include:

(1) *Blades and attachments*. It is important to ensure that the correct blade has been selected. A multipurpose or universal blade offers the greatest flexibility, but specialist tasks may require the use of angle, cushion or straight blades.

(2) *Drawbar pull, engine output and operating weight*. These specifications provide information regarding the machine size and strength. A machine that is too large inhibits manoeuvrability, whilst conversely, a machine that is too small can cost the organisation in terms of reduced productivity and efficiency. The performance specifications of bulldozers range from a drawbar pull of 10,000–215,000 kg, an operating weight of 6.17–132.00 tonnes and an engine output of 50–784 KW.

(3) *Undercarriage configuration*. This broadly relates to the transmission, tracks and sprockets. Most companies offer three main undercarriage configurations, namely (a) *general purpose* for 'normal' environmental conditions (i.e. for working on a range of surfaces from soft ground to rock); (b) *reduced ground pressure* such as the 'swamp track' design where the machine's load is distributed over a greater surface area; and (c) *long* where a high quality of grading performance is required.

Figure 4.2 Tracked loader.

Tracked loaders

The tracked loader is related to the bulldozer but is fitted with a bucket clamshell. The clamshell design opens at the bottom to allow ease of discharge into awaiting dump trucks or other haulage vehicles; it also enables the bucket to grip objects (e.g. tree stumps). Therefore, the tracked loader can dig as well as doze, lift heavy objects, load haulage vehicles and other reciprocals and transport substances and articles around site (Figure 4.2). The mechanics of loading dump trucks is similar to loading with a face shovel tracked 360° excavator since both machines utilise a bucket clamshell.

Excavators

The first excavator was a steam powered shovel manufactured by Otis in the USA, during 1835. Approximately 40 years later, Ruston, Poctor and Burton manufactured the first English steam powered shovel. However, the modern form of hydraulic excavator was conceived during the 1940s following the mechanical engineering revolution inspired by the Bruneri brothers in Italy. Since this latter date excavators have become one of the most versatile and prolific of all plant items. The diversity of this range of machines is vast; they can vary from a 1 tonne 'mini-excavator' to a huge 12,244 tonnes dragline excavator! Similarly, the applications of these machines vary considerably and include foundation excavation, mineral mining, truck loading, utilities and so forth. There are numerous manufacturers of excavators including Caterpillar, JCB, Komatsu, Volvo, Liebherr and Hitachi.

All variants of excavator are operated using joystick controls or levers to manoeuvre the machine arm and slew the machine, whilst foot pedals are used to track the whole machine backwards and forwards. The machine excavation arm forms a skeletal frame and consists of three interconnected parts. These are the boom, the stick (otherwise known as the dipper) and the attachment (normally a bucket). The three parts are connected by bushes and pins (joints) and powered by hydraulic rams (boom, dipper and bucket rams), or steel ropes, that act as the machine's 'muscles'. The machine arm, operator cab and engine and hydraulic compartments are mounted on a slew ring that provides 360° slewing (or swing) capability. This feature ensures that the machine can operate 'unhindered' in relatively confined spaces.

Since there is such a wide choice available, careful consideration should be given to the following when purchasing (or hiring) an excavator:

1. range and availability of attachments (e.g. impact hammers, cutting shears, concrete crushers and so forth);
2. machine cycle times (since these will influence production rates);
3. protective features (audible alarms, falling object protection, convex mirrors, rear view cameras, etc.);
4. engine output and operating weight as indicators of strength and power;
5. running costs to include both fuels and consumables;
6. physical site constraints (e.g. a large machine operating on a small site would improve productivity but it may also increase the probability of accidents occurring).

Excavators are manufactured in three core classifications namely, tracked, wheeled and mini-excavator. These three classifications are now looked at in greater detail.

Tracked excavators

At the heavy end of the spectrum, the 'three phase' electric powered rope excavators still predominate but these are restricted to the largest of opencast excavations in Africa and America. In the mid range, the tracked 'diesel powered' hydraulic excavator is perhaps the most popular and can be seen in use throughout the industrialised economy. This is mainly as a result of the machine's versatility. Similar to the bulldozer, the track type undercarriage allows the excavator to operate in adverse ground conditions. Within this mid range, two broad configurations exist, namely, the face shovel and the backacter.

Face shovel

The *face shovel*, as its name suggests, pushes material away from the machine using a clamshell bucket. Thus, the open bucket faces the material being excavated (refer to Figure 4.3). To operate the face shovel efficiently, a flat, well prepared and compacted substrata is required. This face shovel configuration is particularly suited to operating at the bottom of opencast mines since the bucket literally peels a layer of the material from the bottom of the material pile to the top, in an upward 'cutting' motion. These machines typically weigh between 65 and 800 tonnes, although more specialist 'coal shovels' range between 15 and 25 tonnes.

Backacter

Conversely, the *backacter* machine pulls excavated material 'into' the machine (refer to Figure 4.4). To operate this machine efficiently requires an analysis of the work environment

Figure 4.3 Face shovel tracked 360° excavator.

Source: This figure is reproduced with kind permission from Marubeni-Komatsu Ltd.

Figure 4.4 Backacter tracked 360° excavator.
Source: This figure is reproduced with kind permission from JCB.

to determine any operational constraints that may influence machine performance. For trenches and site clearance operations, the machine should ideally work on a level surface. For production 'earthmoving' operations, the machine would work on a 'bench' which effectively raises the height of the machine above the dump target in order to reduce cycle time.

Three common variants of the backacter hydraulic machine are the long reach, zero tail swing and demolition variants.

(1) *Long reach* variants can be used to clean ditches and rivers. In the latter case, the slewing ring, cab and arm are fitted to specialist barges (tracks and undercarriage are redundant). These then allow the machine to travel water lanes and canals with ease. The only disadvantage with this variant is that an increase in machine arm length normally translates into a reduction in maximum bucket load capability.

(2) *The zero tail swing* option allows the rear of the machine to swing within the confines of its track width, and thus reduces the likelihood of contact with pedestrians or other potential obstructions to the rear of the machine (Figure 4.5). This type of machine has one of the better health and safety design features available, particularly where machines operate in confined spaces.

(3) *Demolition machines* are fitted with falling object protective structures (FOPS). They are often reinforced with wear plates to protect the undercarriage from damage due to demolition debris and a tailgate wear plate to protect the rear structure of the machine from

Figure 4.5 Zero tail tracked 360° excavator.

Source: This figure is reproduced with kind permission from Marubeni-Komatsu Ltd.

Figure 4.6 Concrete crusher attachment.

Figure 4.7 Long reach configuration.

Source: This figure is reproduced with kind permission from Marubeni-Komatsu Ltd.

impact damage (e.g. through contact with debris material such as concrete) when slewing. The tailgate wear plate may also act as a heavy counterweight to balance the machine when lifting heavy loads. In addition to these safety features, demolition machines can also be fitted with a wide range of attachments such as concrete crushers (Figure 4.6), steel shears and grabs of various types. These attachments allow the machine to efficiently dismantle a wide variety of structures. The demolition option may include a long reach configuration (Figure 4.7).

Protective features, such as FOPS and wear plates, can also be fitted to machines operating in other arduous operational conditions, for example, within mining and quarrying or wherever a danger from falling objects exist.

Figure 4.8 Wheeled 360° excavator.

Source: This figure is reproduced with kind permission from Finning (Caterpillar) UK Ltd.

Wheeled excavators

Wheeled excavators are predominantly backacter machines and are more fondly referred to as 'rubber ducks' because of the nodding action of the boom as they travel. They are similar to operate as their tracked counterparts but include rubber tyres instead of tracks, hence their name (refer to Figure 4.8). Tyres enable the machine to travel on surfaces that would perish under the stress of steel tracks, for example, tarmacadam or other road surfacing materials. The wheeled machine sometimes includes a dozer blade that is incorporated into the design and can be used to grade substrata material, push excavated material back into trenches, and act as a stabilizer when excavating.

A variant of the wheeled machine can be found in the rail industry and is named the road railer (Figure 4.9). These machines are simply fitted with rail wheels (in addition to tyres) thereby enabling them to travel unrestricted distances along the rail network. In such circumstances, these machines are typically used to perform track and embankment repair and maintenance works.

Mini-excavators

Richard Smalley (Lincolnshire, England) is widely acknowledged as the genius behind the development of the 360° mini-excavator (*c.* early 1960s). Initial mini-excavator models were

Figure 4.9 Road railer.

Source: This figure is reproduced with kind permission from Marubeni-Komatsu Ltd.

Figure 4.10 360° mini-excavator.

Source: This figure is reproduced with kind permission from JCB.

manufactured on 'stilts' and they moved forward by 'dragging' themselves along, that is, the backacter machine arm was stretched out, lowered to the ground and used to pull the mini-excavator forward. This original design is still used today as a purpose-built gravedigger. This particular type of machine also incorporates telescopic stabilisers that help prevent machine turnover during excavation. Later models were manufactured on wheels for use on tarmacadam road and other hard surfaces. Japanese manufacturers then further developed the mini-excavator by including tracks and a slew drive, with these tracks being constructed from either steel or hard-wearing rubber. Rubber tracks are a good compromise since they enable the machine to work both on and off highways (and other softer surface materials) (Figure 4.10). Akin to the wheeled excavator the mini-excavator can also include a dozer blade.

Applications of the mini-excavator include small domestic and industrial construction work (trench excavation, site strip and so forth), other work where space is at a premium (i.e. where a mid range '12–30 tonne' excavator would not fit), landscaping and utility services (e.g. gas and water). A combination of low capital/running costs, high production and compact design has ensured that the mini-excavator has replaced manual excavation and traditional materials handling methods for most small projects. World sales are dominated by Komatsu and Hitachi, although JCB, Kubota, Volvo, Hanix, Bobcat and Case also manufacturer 'minis'. In 1980, UK mini-excavator sales were a mere 200 whilst in 1998 over 4,000 were sold. This almost exponential growth in sales clearly demonstrates the mini-excavator's increasing utilisation within UK construction.

Compactor rollers

The road, rail, construction, mining and waste industries all heavily rely upon the use and reuse of excavated substrata and aggregates to backfill trenches and foundations. However, following excavation, such materials are bulked by voids of air that leave them susceptible to settlement and distortion after placement. To reduce the likelihood of this type of failure, the substrata density must be artificially increased by compaction methods, to increase substrata strength and reduce its compressibility. The amount of achievable compaction depends upon the soil's chemical and physical properties and the compaction forces exerted upon it. The variance in soil types and desired compaction values have inevitably led to the manufacture of various types of compacting equipment. The traditional means of compacting ground was achieved through the use of a towed roller attached to a crawler loader, tractor or bulldozer. Although the towed roller is still used, a range of specialist compactors is now available; these are the drum roller, sheep's foot roller, vibrating roller and deadweight roller. Manufacturers of compactor rollers include Caterpillar, Komatsu and Bomag.

Single and double drum rollers

The single, smooth-drum pedestrian roller is the smallest of the drum rollers. The machine's diesel engine powers a heavy vibrating drum to compact soil, expel water and drive the machine forward. The vibration of the drum increases the effective weight of the roller by up to ten times its dead weight. For this small roller, the steering system is simplistic and is achieved via the use of a handlebar which allows forward and reverse movement. The single drum roller is ideally suited to pavement and small works construction where limited space prevents the use of a larger machine.

The 'ride on' double smooth-drum roller is a heavier machine and is articulated in the middle to improve steering (Figure 4.11). Both drums vibrate to give maximum compaction. Typical applications include wide trenches, pavements and foundations.

Figure 4.11 Ride on double smooth-drum roller.
Source: This figure is reproduced with kind permission from Bomag.

Sheep's foot roller

The sheep's foot roller is a specialist variant of ground compactor roller designed to work primarily on poor ground conditions, such as land fill applications. The machine has a front mounted dozer blade to help level the material before it is compacted and a double vibrating roller drum equipped with sheep's foot metal spikes (Figure 4.12). These spikes replicate the tamping action of a sheep's hooves and drive hard into the surface material to achieve extra compaction. The amount of compaction depends largely upon the weight of the roller and the contact area of the metal spikes. The smaller the area in contact with the ground, the greater the stress (compaction) imposed upon it (stress = load/area). This is the converse situation to that of tracks which distribute the load of a machine over a greater area thus providing floatation for a tracked machine operating on poor ground conditions.

Self propelled vibrating roller and deadweight roller

For larger pavement construction (e.g. airports and highways), the self propelled vibrating roller and the deadweight roller are often used in conjunction. The self-propelled roller is articulated for easy steering and consists of one smooth drum at the front of the machine and two, almost smooth, rubber tyres at the rear (Figure 4.13). The machine is designed to compact stone and other aggregates to form a suitable bed for bituminous materials and is much larger and heavier than its single and double drum counterparts.

Figure 4.12 Sheep's foot roller.

Source: This figure is reproduced with kind permission from Finning (Caterpillar) UK Ltd.

Once the sub-base has been sufficiently prepared, coarse and finish surface coats are then laid and compacted using the deadweight roller (Figure 4.14). The wheel base configuration for this machine consists of two slim rear rollers and one wide front roller that overlaps the rear rollers to ensure that ridges in the top coat are not created. Water is sprinkled onto the rollers during use to ensure that the asphalt or tarmacadam does not stick to them. Some machines also include a cutting wheel that maintains a straight edge ready for an adjacent bituminous strip to be laid (if required).

Other compaction equipment

The pneumatic-tyred roller has two sets of pneumatic tyres which press road chippings into a tar binder laid on road surfaces. Generally the machine is lighter than other ride on rollers

Figure 4.13 Self propelled vibrating roller.

Figure 4.14 Deadweight roller.

Figure 4.15 Vibrating compactor plates.

Source: This figure is reproduced with kind permission from Bomag.

Figure 4.16 Vibrating tampers.

Source: This figure is reproduced with kind permission from Bomag.

(e.g. the self propelled vibrating roller) and its pneumatic tyres ensure that the chippings are not crushed to dust.

Other types of compaction equipment include vibrating compactor plates (Figure 4.15) and vibration tampers for smaller projects (such as service and domestic foundation trenches) (Figure 4.16). The compactor plate is sometimes referred to as a 'whacker plate' after the manufacturer in the same way that vacuum cleaners are incorrectly called Hoovers! In addition to compacting foundation and services base material, the steel plate can also be covered with a purpose designed rubber mat, carpet or hessian and used to compact block paved areas. The inclusion of a tough material cover over the steel compactor plate ensures a reduction in pattern marking on the blocks that occurs as a result of grit or stone becoming trapped between the steel plate and the blocks.

Wheeled 'tractor' backhoe loaders

The wheeled backhoe loader is one of the most popular of off-highway plant and equipment items available for the owner operator. The machine combines the features of the wheeled loader (bucket clamshell) and excavator (backacter machine arm with up to 180° slew capability) and because of this, it has great versatility and manoeuvrability both on and off highways. For many smaller businesses, the wheeled backhoe loader is the first and often only machine purchased.

Within the UK, this particular machine is synonymous with JCB although the exact origin of the first backhoe loader is vehemently contended amongst plant manufacturers (namely, Case and JCB). However, the referral to a 'JCB' as opposed to a specific model of machine provides strong evidence of JCB's prominence within the plant and equipment industry. In 1953, Mr Joseph Ceril Bamford amalgamated the design principles of a lightweight backhoe manufactured by Eik Hauskins Co., a Ford Major Tractor and the wheeled loader face shovel. This new 'revolutionary' machine, called the MK1, was sold to the agriculture, civil engineering and construction industries commencing in 1954. Since then, the broad shape and functionality of the machine has hardly changed. Engineering alterations have been made but these have tended to focus upon improvements to operational performance, efficiency, power and operator comfort.

This machine's basic design configuration has provided great advantages over mechanical rivals (Ober, 1999). Essentially, the machine anatomy consists of four key components, which are:

1 a four wheel drive to enable good manoeuvrability in even the most arduous of environments;
2 a face shovel to carry aggregates or crops, or to strip, for example, top soil;
3 a backacter to excavate trenches, move spoil, grade embankments or dredge rivers/pools;
4 stabilisers to ensure minimal machine movement whilst the backacter is being utilised (Figure 4.17).

Both the face shovel and backacter can be fitted with a variety of tools, for example, impact hammers for breaking hard surfaces, road sweeping brushes for agriculture and so forth.

Figure 4.17 Wheeled tractor backhoe loaders.
Source: This figure is reproduced with kind permission from JCB.

World market leaders of this product today include JCB, Case and Caterpillar. However, sales of the machine have declined as a result of the development of other complementary machines, namely the mini-excavator and the telehandler. To combat this decline, mechanical design engineers have developed hybrid 360° wheeled backhoe loaders that are smaller and more versatile. This development integrates the skid steer loader, 360° mini-excavator, and backhoe loader into one machine.

Forklift trucks and telehandlers

The American Society of Mechanical Engineers (ASME) defines a powered industrial truck as a 'mobile, power-propelled truck used to carry, push, pull, lift, stack, or tier materials'. Powered industrial trucks are also commonly known as forklifts, pallet trucks, tele-truck, rider trucks, forktrucks, or lifttrucks (Figure 4.18). The definition cited, however, equally applies to all forms of off-highway forklift, since it conveys the machine's fundamental purpose and functionality, that is, materials handling. The forklift machine eliminates the need for hoists or even craneage on low rise developments.

The packaging of materials and prefabrication of components has stimulated the need for greater transportation of these goods. To satisfy this need, two types of forklift truck have evolved and now predominate in construction and agriculture. These are the rough terrain telescopic handler (often shortened to telehandler) and the rough terrain fork lift (masked) truck.

Figure 4.18 Tele-truck (forklift truck).

Source: This figure is reproduced with kind permission from JCB.

Figure 4.19 Rough terrain telescopic handler.

Source: This figure is reproduced with kind permission from JCB.

Rough terrain telescopic handler

Rough terrain telescopic handlers are manufactured to a variety of weight capacities (typically 1.5–5.0 tonnes) to suit both light and heavy loads (Figure 4.19). To provide extra stability when loading at heights, out-riggers (otherwise known as stabilisers) are sometimes incorporated into the machine's design. Some practitioners argue that this feature is unnecessary when the machine is equipped with a safe load indicator. Indeed, the risk of overturning the machine is higher when it is in transit and loaded. To reduce this type of transit accident, loads should always be carried low to the ground and travel speeds kept to a minimum.

The outstanding feature of a telescopic handler is the incorporation of an extendable arm with a capability similar to the modern mobile crane. Some machines possess an 18.0-m lift 'reach' height. In addition, the machine's arm can be fitted with a wide range of attachments (e.g. road brushes, forks, hooks and so forth). This particular 'multi-capability' feature of the machine enables it to perform a variety of tasks, such as:

1 lifting palettes of materials onto scaffolding;
2 acting as a mobile crane, therefore reducing the need for a labour intensive workforce or hiring/purchasing of additional plant items (i.e. mobile cranes, rigid forklift trucks and so on);
3 loading from delivery vehicles direct to the point of usage, which reduces the need for storage space on site; a much valued commodity;
4 stacking materials in stockpiles in order to maximise available storage space.

The telescopic handler may also use a rough terrain four-wheeled 'power train' drive design, making it ideal for agricultural as well as construction use. Hence, 40 per cent of total sales of this type of machine are within the agricultural industry.

Within the UK, the Giraffe, manufactured by Liner, was probably the first formal telescopic handler, although this particular manufacturer has recently ceased trading. Currently, the main manufacturers supplying UK industry are JCB, Manitou and Merlo. JCB and Manitou are currently world leaders in telescopic handlers, as measured by the number of items sold. Sales figures for the UK construction industry are currently estimated to be around 4,000–5,000 machines, a growth from fewer than 200 when records of machine sales were first recorded in 1980. This rapid growth in machine sales provides evidence of the machine's success within UK construction.

Rough terrain forklift 'masted' truck

The rough terrain forklift truck represents a more traditional machine and has been widely used since the 1960s. Machine anatomy is similar to the telehandler but the main difference is the inclusion of a vertical telescopic mast equipped with forks (Figure 4.20).

It is important that both the rough terrain telescopic handler and the rough terrain fork lift 'masted' truck are equipped with FOPS due to the nature of their typical application, that is, materials distribution and/or handling at heights.

Dump trucks

For earthmoving, civil engineering and mining activities, the transportation of rock, earth and assorted minerals is an essential task. Dump trucks represent the most widely utilised means by which to haul material over a given distance. They range from the small construction

Figure 4.20 Rough terrain fork lift 'masked' truck.

Source: This figure is reproduced with kind permission from JCB.

site dumper, to the huge 'rigid' haulers found in the mining industry. Dump trucks are more efficient than any other off-highway plant at hauling material over long distances. However, these machines are reliant upon excavators or wheeled loaders to load them.

Site dump truck

The site dumper is a popular machine, seen on almost every small to medium size earthmoving project since it can transport a wide range of materials (e.g. aggregates, bricks and cement). Similarly to other off-highway machines, the site dumper is manufactured in either a tracked or wheeled configuration. The wheeled variety is by far the most commonly used (Figure 4.21) and can incorporate two- or four-wheel drive for either easy or arduous ground conditions, respectively. The site dump truck's tipping action can also vary between the standard forward dump action or the side-tip action of skips mounted on a turntable. Ultimately, the type of machine selected depends upon its intended application. Forward tipping, four-wheel drive machines are seen throughout industry, whilst side tip machines are more useful for working in restricted areas (such as highways maintenance).

Current manufacturers of site dumpers include Barford, Benford, Thwaites, Ausa and Lifton. Payload capacities typically range from 0.50 to 6.00 tonnes with body capacities of 0.18–3.2 m^3. Travel speeds vary between 4.3 and 32 km/h.

Articulated dump truck

The world's first articulated dump truck (ADT) was developed by Volvo in 1957 and was called the 'Moon Rocket'. The concept underlying the design of ADTs is the use of a connection between the front and rear frame and an all wheel (four or six) drive (Figure 4.22). These

Figure 4.21 Site dump truck.

Figure 4.22 Articulated dump truck.

Source: This figure is reproduced with kind permission from Finning (Caterpillar) UK Ltd.

features enable the front and rear of the machine to move independently of each other in two planes, thus reducing stresses induced onto the structural frame when transporting material over undulating ground conditions. A rear-tipping action is used to deposit material collected.

In many instances, scrapers have been replaced by ADTs, principally because ADTs have similar productivities but are more economical to operate, even though they need other machines to load them. Current manufacturers of ADTs include Aveling Barford, JCB, Caterpillar, Komatsu, Terex, Bell, O&K, Moxy, Volvo and Thwaites. Payload capacities typically range from 8.5 to 37 tonnes and heaped capacities from 4.8 to 22.5 m^3. Travel speed ranges from 25 to 56 km/h.

Rigid dump truck

In 1933, Euclid introduced the first large-capacity 'self powered' off-highway rigid dump truck for the construction, mining and quarrying industries. Rigid machines have a rear tipping action and include a reinforced body that extends over the cab to protect the driver from falling debris during loading (Figure 4.23).

The sheer size, machine weight and payload of rigid dump trucks dictates that haul roads must be hard and stable. If they are not, the truck's structural frame may twist or crack, thereby incurring further expenditure and/or causing accidents. Standard practice therefore

Figure 4.23 Rigid dump truck.

Source: This figure is reproduced with kind permission from Finning (Caterpillar) UK Ltd.

includes the application of other plant such as bulldozers and compactors to ensure that haul roads are suitable for rigid dumper use.

Manufacturers who presently include rigid dump trucks in their product line are: Liebherr, Case, Terex, Aveling Barford, Euclid Hitachi, Caterpillar and Komatsu. Payload capacities typically range from 10.5 to 350 tonnes with heaped capacities of 5–183 m^3. Travel speed in forward varies considerably between 32 and 68 km/h.

Cranes

Prefabrication of large manufactured construction components and pre-packaging of construction materials has fuelled the need for machines that lift, lower and transport heavy loads quickly and efficiently. Cranes have evolved to meet these needs and are used widely for construction, maintenance, bridging and re-supply activities (shipping, rail, agriculture and so forth). The origins of cranes can be traced back to 1556 but the first 'true cranes' were developed by the Victorians (Johnson, 1998). These steam powered machines proved invaluable to the then mighty British Empire Naval Fleet who used cranes for stocking ships with materials and resources (ibid.).

In the modern era, two broad classifications of crane have evolved and now predominate, namely tower cranes and mobile hydraulic cranes. There are also some 'hybrid' varieties of crane such as the mobile tower crane.

Tower cranes

For high-rise construction and civil engineering works, tower cranes are a vital resource. Although there are various categories of tower crane available, the *horizontal, luffing* and *articulated jib* varieties are most common. Each variety is designed to resist overturning when lifting and torsion from side loads acting on the boom, for example, wind loads.

Horizontal jib

The *horizontal jib* is fixed, constructed using a lattice structure and is divided into two sections. Both sections are mounted on a vertical standing mast that is firmly anchored to the ground via a sturdy concrete foundation and hold down bolts. For the first section, reinforced steel cables extend from a winch to the jib trolley and hook block. The hook block is attached to the jib to allow loads to move across the length of it. The second section is much shorter and incorporates a counterweight to stabilise the crane when lifting loads. To manoeuvre a given load (right or left and through 360°), a slewing ring is located at the intersection between the top of the tower and the jib. The operator's cab is often located on top of the slew ring such that a good, clear view of the load and ground is achieved (Figure 4.24). A requirement of the horizontal jib crane is that there is an unrestricted horizontal clearance; this enables the crane to bring loads close into the vertical mast ready for unloading.

Luffing and articulated jib

A *luffing jib* tower crane consists of a lattice-framed vertically standing mast mounted on a sturdy turntable that is powered by electricity. For this type of tower crane the boom is not fixed and hence can be raised clear of any site obstructions (e.g. on confined sites that may include various obstacles). The articulated jib has similar functionality to the luffing jib counterpart but as its name suggests, the jib is articulated so as to enable even greater versatility when handling loads.

Figure 4.24 Horizontal jib tower crane.

Source: This figure is reproduced with kind permission from Liebherr (Great Britain) Ltd.

Other configurations

There are various other types of tower crane configuration available and these include:

1 *Rail mounted crane* (otherwise known as the travelling crane): This type of crane is mounted on tracks and is particularly useful on sites where increased load distribution capabilities are required.
2 *Tied-in tower crane*: For structures above 100 m high, a tower crane tied-in to the structure, at intervals recommended by the manufacturer, is used. By tying into the structure, greater resistance to overturning and torsion is provided whilst simultaneously ensuring that greater heights are reached.
3 *Climbing crane*: On sites where external space is limited (e.g. in city centres where sky scrapers predominate), the climbing crane is used. The position of the crane is located on the inside of the structure and is secured to key structural components. As each consecutive floor of the structure is completed the crane sheds its fixed base and climbs to the next floor using either winches or hydraulic jacks.

Figure 4.25 Mobile 'telescopic' crane.

Source: This figure is reproduced with kind permission from Liebherr (Great Britain) Ltd.

Figure 4.26 Mobile 'tracked' crane.

Source: This figure is reproduced with kind permission from Liebherr (Great Britain) Ltd.

Mobile cranes

The mobile hydraulic group of cranes includes crawler, all terrain, truck mounted, rough terrain and city cranes. The mobility of this group gives them a great advantage over tower cranes, particularly where lifting activities are spread over a wide geographical area. Inclusion of tyres means that the city, truck mounted, rough terrain and all terrain cranes can travel on the highways between various locations (Figure 4.25). The all terrain and rough terrain cranes (as their name indicates) can also travel off-highway thus enabling penetration into remote locations. Load capacities vary but tend to range between 30 and 800 tonnes, although the Riga AMK 1000 telescopic crane has a 1,000 tonne lift capacity.

For special lifting operations, lattice booms fitted to crawler tracks have been developed (Figure 4.26). This combination of simple but effective design allows the machine to operate on poor ground whilst simultaneously lifting heavier loads (usually between 250 and 1,200 tonnes). The world's biggest mobile 'tracked' crane was built by Demag and can lift 2,000 tonnes. Crawler cranes also offer good mobility but must be to some extent dismantled and transported on low loader trailers between projects. Additional costs are therefore incurred in terms of time and other resources needed to erect the crane, along with transportation costs. Manufacturers of these machines include Tadano-Faun, Kato, Liebherr and Grove. Lift capacities range from 3 to 1,600 tonnes.

Wheeled loaders and integrated tool carriers

Wheel loaders and tool carriers are relatively new machines when compared to other, more familiar off-highway plant and equipment (Figure 4.27). The original machine was

Figure 4.27 Wheeled loader.

Source: This figure is reproduced with kind permission from Finning (Caterpillar) UK Ltd.

conceived in the agricultural industry during the 1920s when tractors were fitted with a loading shovel. However, the first self-contained, two wheel drive, rubber-tyre wheeled loader was introduced by Frank G. Hough Co. much later in 1939. Early engineering designs aimed to produce a face shovel bucket attachment to a machine that could re-handle light agricultural materials. Such attachments are still available today. However, in later years the 'front-end' machine became far sturdier and suitable for carrying heavier payloads; it also had higher breakout forces.

This front-end machine predominates today in the aggregates industry and is used widely in construction and mining. It is designed specifically as a digging and loading machine, although, some of the larger machines have quite recently been adapted to serve as rock shovels in the mining sector.

There are many wheeled loader manufacturers worldwide but the prominent manufacturers include Komatsu, Volvo, Caterpillar, JCB, Liebherr, Fiat-Hitachi, O&K, Case and New Holland. Machines are either rigid or articulated but the majority are articulated. Key machine specification variables include shovel capacity (0.08–20 m^3), max speed (7–35.2 km/h), breakout force (between 7 and 274 kN) and operating weight (0.65–205.20 tonnes).

Skid steer loaders

The skid steer loader (also known as the compact utility machine) is a four-wheel drive vehicle with rigid axles. It manoeuvres by powering the wheels (or tracks) on one side of the machine whilst applying brakes to the other. This independent movement of the wheels on one side allows it to rotate through 360° within its length similar to a military combat tank. Typical machines (such as the Caterpillar models) have no transmission. Instead, they are powered by the use of hydraulic pumps and motors (one motor for each side of the machine). Each motor connects to a sprocket that is linked by two chains to each wheel. The sprockets

Figure 4.28 Skid steer loader.

Source: This figure is reproduced with kind permission from JCB.

and chains thus distribute the power from the hydraulic motor and increase the torque at the wheels by providing a gear reduction. Up to four hydraulic pumps are used; two for drive power, one for lift/work 'attachment' tool power and one for circulating hydraulic oil through filters to provide pressure for pilot controls. This innovative engineering design ensures that the machine does not stall during utilisation (Figure 4.28).

The first 'true' skid steer loader was probably developed by Bobcat (*nee* Melroe Manufacturing Company) in 1960. This machine, named the M-400 was initially developed for the agriculture industry, but soon revolutionised the wider compact materials handling sector. The name 'Bobcat' was reputedly used in honour of the prairie animal whose 'tough, quick and agile' attributes resemble those of the machine. New Holland's engineering designer, Larry Halls, also designed a competitor called the Super Boom (affectionately called the Flintstone buggy) in 1960.

When first developed, general industry was not quick to grasp the inherent functionality of these robust, compact and highly adaptable machines, so sales of the skid steer loader failed to grow until the early 1980s. Since then, the machine has become an essential item for any company engaged in materials handling activities. Principally, this reliance is due to the considerable range of attachments available; up to 40 individual work tools at present with future development potential. Some of these items include augers, material handling forks, trenchers, demolition tools, buckets, snowblowers and so forth. This diversity of attachments gives the machine unrivalled switching capability such that a number of tasks can be completed by one single item. In turn, the future of skid steer loaders in wider industry has been secured, for example, in agriculture, construction, civil engineering, shipping and demolition. Because of the diversity of operational requirements, the machine specifications also vary considerably with operational weight varying between 0.53 and 4.5 tonnes.

Part II

Investment, procurement and business management

Chapter 5

Economic comparisons of equipment alternatives

Principles of economic comparisons

The basic approach to economic comparisons is to assemble all the costs relating to one course of action and all the costs relating to the alternative course of action and to compare them. The assembling of the costs must be in such a way that the two are comparable. It is the difficulty of ensuring that the assembled 'packages' of costs are comparable that requires the calculation of either present worth, or value, of two proposals or the equivalent annual cost of the two proposals. Of these techniques of comparison, present worth is more commonly used. Both present worth and equivalent annual costs require an interest rate which is taken to represent the value of money to the investor. That is, it represents the interest the investor could receive elsewhere.

Comparisons not involving interest rates are very common in short-term schemes – that is, schemes of less than one year. Company staff are continually and almost subconsciously undertaking economic comparisons without interest in the calculations. Such economic comparisons include comparing the hire rate for different cranes, or the hire rate for different excavators, or the hire rate for an excavator from an external equipment hire company with the hire rate from the enterprise's own equipment division. Comparing the cost of, say, hand excavation with the cost of using an excavator is another economic comparison. All these comparisons refer to operations with relatively short time periods of a few weeks or a few months and the effect of interest is not significant. Thus, the comparisons are valid and easy to make on an equitable basis. The comparisons become more difficult when the operations or schemes to be compared last a few years or more, when the effect of interest becomes significant and needs to be included in the calculation. The difficulty is assembling the various costs into 'packages' that can be compared for these longer-duration operations that require 'present worth' or 'equivalent annual costs'. Some examples of longer-duration operations or schemes are equipment required for quarrying, open-cast mining or concrete production.

All the examples in this section are based on cash flows that have been estimated at present or year zero prices without taking inflation into account. The interest and time relationships used in this chapter are explained in the Appendix, which includes examples of interest tables for interest rates of 10 and 15 per cent.

Present worth

Present worth comparisons are used to compare two or more schemes where the equipment chosen for each scheme leads to different capital investment and different running costs.

Essentially, present worth enables the trade-off between capital investment and future running costs to be compared. Example 5.1, which includes the capital cost of buying equipment and the running costs of operating the equipment, illustrates the comparison. All the estimates used in this comparison are at present-day (i.e. year zero) prices.

Example 5.1

Assuming an interest rate of 15 per cent:

	Proposal 1	Proposal 2
Capital cost of equipment	£8,500	£9,500
Annual running costs	£1,750	£1,500
Life	5 years	5 years

The present worth of Proposal 1 is £14,366 and of Proposal 2 is £14,528. The calculations are as follows:

Proposal 1

$$\begin{aligned}\text{Present worth} &= £8{,}500 + (£1{,}750 \times 3.352)\\ &= £8{,}500 + £5{,}866\\ &= £14{,}366\end{aligned}$$

Proposal 2

$$\begin{aligned}\text{Present worth} &= £9{,}500 + (£1{,}500 \times 3.352)\\ &= £9{,}500 + £5{,}028\\ &= £14{,}528\end{aligned}$$

The capital sums are already in year zero and need no further manipulation. The running costs of £1,500 and £1,750 each year need to be converted to present worth or capital sums. The factor used for this conversion is the uniform series present worth factor. This factor, 3.352, has been taken from the tables in the Appendix.

The present worth of Proposal 1 is £14,366 and the present worth of Proposal 2 is £14,528. Thus, Proposal 1 is the more economic. The present worth of Proposal 1, £14,366, represents enough money to buy the equipment item at a cost of £8,500 and investing the remainder at 15 per cent is enough to produce £1,750 each year for the next five years. Thus £14,366 is the amount required now to meet all the requirements of Proposal 1. Similarly, the present worth of Proposal 2, £14,528, is the amount required now to meet all the requirements of Proposal 2. Since the present worth of Proposal 1 is the smaller and both proposals would be compared only if the two items of equipment were capable of doing the same tasks, then the one with the least cost, that is, least present worth, is the most economic.

An alternative way of considering this comparison is to examine the differences between the capital costs and the running costs. Proposal 1 has £1,000 less capital but requires £250 more running costs each year. Thus, the extra capital of £1,000 involved in Proposal 2 can be seen to be buying £250 of savings in the running costs. The question as to which of the two schemes is the more economic could be restated as follows: would the £1,000 of extra investment be better used in saving £250 in running costs or earning 15 per cent if invested elsewhere? £1,000 invested at 15 per cent for five years would give an income of £298.31, calculated as £1,000 × 0.29831 = £298.31; where 0.29831 is the capital recovery factor, taken from the tables in the appendix. Thus, the return on the investment is better than the saving in the running costs. Therefore, Proposal 1, the smaller of the capital investments, is the more economic proposal.

This present worth comparison is valid so long as the lives of the two proposals are the same. This is usually the case in comparing equipment items. One exception is comparing the cost of keeping an item of equipment for one year with that of keeping it for two, three or four years. In such cases the lives are different and require different treatment. This is explained later, in the section dealing with replacement.

The example presented is the use of present worth in a simple case where there are only capital and running costs and where the running costs were uniform – that is, the same each year. The principles of using present worth are the same even when the cash flow becomes more complicated. The next example shows the running costs varying each year in order to reflect the increasing costs incurred as the equipment grows older. Also included in the next example is a resale value of £4,000 occurring in the last year. The resale value is a return of money to the investor and therefore carries a different sign from the capital and running costs, which are outflows of money. Again all the cash flows are estimated at present prices.

Example 5.2

Assuming an interest rate of 15 per cent:

Year	Cash flow for the purchase and resale of an item of equipment (£)
0	−8,500
1	−1,750
2	−1,850
3	−2,000
4	−2,200
5	−2,500 + 4,000

The present worth of these cash flows is £13,247.72, calculated as shown in Table 5.1 at an interest rate of 15 per cent. This present worth can now be compared with the present worth for an alternative proposal.

Table 5.1 Calculations of present worth for Example 5.2

Year	*Cash Flow* (£)	×	*Present worth factor (15%)*	=	*Present worth* (£)
0	−8,500		1.0		−8,500.00
1	−1,750		0.86956		−1,521.73
2	−1,850		0.75614		−1,398.87
3	−2,000		0.65751		−1,315.02
4	−2,200		0.57175		−1,257.85
5	−2,500		0.49717		−1,242.93
5	+4,000		0.49717		+1,988.68
			Total present worth		−13,247.72

The present worth factors are taken from the tables in the Appendix. Because the annual sums are varying in this case, the uniform series present worth factor cannot be used and the individual present worth factor for a lump sum, $1/(1+i)^n$, is used instead, where i is the interest rate and n is the number of years. This increases the arithmetic involved but is unavoidable when dealing with varying annual cash flows.

The treatment of the £4,000 resale value shown here is to simply add it into the present worth, taking account of the different sign. It may be more acceptable to deduct its present worth for the initial capital:

Initial capital invested	= £8,500.00
Present worth of resale = £4,000 × 0.49717	= £1,988.68
Adjusted capital invested	= £6,511.32

The total present worth calculated after adjusting the capital invested in this way to take account of the resale value will be the same as the example given.

Equivalent annual costs

An alternative to present worth comparison is comparing proposals on the basis of the equivalent annual costs. Whereas present worth converts all future running costs to a present worth or capital sum, equivalent annual costs convert the capital sums to an annual cost. Equivalent annual cost comparisons achieve the same as present worth comparisons and, like present worth comparisons, are essentially evaluating the trade-off between capital and running costs. Example 5.3 uses the same cash flows estimated at present prices as Example 5.1.

Example 5.3

Assuming an interest rate of 15 per cent.

	Proposal 1	Proposal 2
Capital cost of equipment	£8,500	£9,500
Annual running costs	£1,750	£1,500
Life	5 years	5 years

The equivalent annual cost of Proposal 1 is £4,285.64 and the equivalent annual cost of Proposal 2 is £4,333.95. The calculations are as follows:

Equivalent annual cost of Proposal 1 = £1,750 + (£8,500 × 0.29831)
= £1,750 + £2,535.64
= £4,285.64

Equivalent annual cost of Proposal 2 = £1,500 + (£9,500 × 0.29831)
= £1,500 + £2,833.95
= £4,333.95

The annual running costs of £1,750 and £1,500 do not need further manipulation. The capital costs of £8,500 and £9,500 need to be converted to annual costs. The factor used for this conversion is the capital recovery factor, 0.29831, and has been taken from the tables in the Appendix.

The equivalent annual cost (EAC) of Proposal 1 is £4,285.64 and the EAC of Proposal 2 is £4,333.95. Thus, Proposal 1 is the more economic, as was found by the present worth comparison. The EAC represents the annual cost of owning and operating the item of equipment. This is made up of £1,750, representing the annual running cost, and £2,535.64, representing the annual cost of the capital investment. This is calculated on the basis that if £8,500 were invested at an interest rate of 15 per cent, an income of £2,535.64 could be taken each year for the next five years. This income would be made up of the interest earned plus the original capital. At the end of the first year £1,275 of interest would be earned on the £8,500 of capital invested. Thus, if an income of £2,535.64 were taken, this would be made up of £1,275 of interest plus £1,260.64 of capital, leaving £7,239.36 of capital. At the end of the second year the £7,239.36 of capital would earn £1,085.90 of interest. The income of £2,535.64 would be made up of the £1,085.90 of interest plus £1,449.74 of capital, leaving £5,789.62. The interest earned in the third year would be £868.44 and the income of £2,535.64 would be made up of the £868.44 of interest plus £1,667.19 of capital, leaving £4,122.43 capital. In the fourth year the interest earned would be £618.36 and the income of £2,535.64 would be made up of the £618.36 interest plus £1,917.27 of capital, leaving £2,205.16. In the fifth year the interest earned would be £330.77 and the income would be made up of the £330.77 interest plus £2,204.86 capital, leaving the account exhausted and no capital (on the basis of the calculations presented here, £0.30 of capital would remain but this is due simply to rounding errors in the calculation). Thus, if the £8,500 were invested at 15 per cent, an income of £2,535.64 could be taken each year for five years. However, since the £8,500 was not invested in such an account but used to purchase the item of equipment, the investor is deprived of the income of £2,535.645 and therefore this can be regarded as equivalent to the annual cost of owning the equipment item.

This example shows the use of EAC where only capital and running costs are considered and the running costs are uniform. The use of EACs becomes more difficult when running costs vary from year to year. In such cases it is necessary to convert the varying running costs to a capital cost before converting them back to an EAC. The next example, using the same cash flows as Example 5.2, illustrates this difficulty. As in Example 5.2, the cash flows are estimated by use of present prices.

Example 5.4

Given that the interest rate is 15 per cent.

Year	Cash flows for the purchase, operating and disposing of an item of equipment (£)
0	−8,500
1	−1,750
2	−1,850
3	−2,000
4	−2,200
5	−2,500 + 4,000

The present worth of the resale, that is, £1,988.68 and the purchase price less resale, that is, the adjusted capital invested, is £6,511.32, as explained in Example 5.2.

The EAC of the capital invested is £1,942.39, calculated as follows, taking the capital recovery factor for five years, 0.29831, from tables in the Appendix:

$$\text{EAC} = £6{,}511.32 \times 0.29831 = £1{,}942.39$$

The present worth of the running costs are £6,736.40, summed from the calculations in Example 5.2 and the EAC of the present worth of these running costs is £2,009.54, calculated as follows, taking the capital recovery factor for five years, 0.29831, from tables in the Appendix:

$$\text{EAC} = £6{,}736.40 \times 0.29831 = £2{,}009.54$$

Thus, the total EAC for all the cash flows – the purchase price less resale value and running costs – is £3,951.93. This is made up of £2,009.54, representing the running costs, and £1,942.39, representing the purchase price less resale. Thus, the EAC of £3,951.93 can be used for comparison with similarly calculated EACs for alternative proposals.

By converting the varying annual running costs to a present worth and then converting this present worth to an EAC, the running costs can be distributed uniformly over each year. This procedure overcomes the difficulty with varying running costs. However, this procedure also illustrates why the use of present worth as a basis for comparison is more common. To arrive at an EAC, the cash flows had first to be converted to a present worth. Therefore, it is easier to perform the comparison on the basis of present worth rather than involve the extra calculation of producing EACs.

However, the EAC method does not require the levels of proposals under comparison to be equal, as is required in a present worth comparison. The reason is that the sum calculated, the EAC, refers to one year and the EAC of any alternative scheme also refers to one year. The process of calculating EACs takes account of the duration of the equipment item and produces annual costs which can be compared with annual costs for other proposals. An example of this is comparing the cost of keeping an item of equipment for one, two, three, four or five years, thereby determining the best replacement age. The costs of keeping an item of equipment for different periods are, in effect, different proposals which have different lives and therefore present difficulties in comparison on the basis of present worth.

Another use of EACs is to convert the purchase price and running costs to annual costs that can then be compared with the cost of hiring. The same device of converting the capital costs to annual costs is one way of determining the capital element in a hire rate when determining what the economic hire rate should be.

Economic comparisons and inflation

In the previous section all the examples were based on cash flows that were estimated at present or year zero prices. The effect of inflation was not taken into account. In Examples 5.2 and 5.4 the annual costs increased as a result of incurring increasing running costs as the equipment grew older and not the effect of inflation. A simple illustration of inflation is that, if in year zero a 'basket of goods' cost £100 and in year one the same basket of goods cost

£110, then inflation of 10 per cent has occurred. In other words, inflation causes more money to be paid out for the same goods. As individual items change price, or inflate, at different rates, the basket of goods concept is used to create indicators of the average price movements or inflation. The most commonly known 'basket of goods' is that used to calculate the retail price index.

The approach of estimating cash flows for purchase price, running costs and resale at present or year zero figures and then including the adjustments for inflation is frequently used, because it separates the difficulties of estimating the cash flows resulting from the selection of the equipment item and its technical capabilities from the vagaries of inflation. This separation also allows different inflation assumptions to be made and evaluated without disturbing the underlying estimates.

The following methods explain the means by which the inflation assumptions can be incorporated into economic comparisons. The explanations are based on the cash flows for Examples 5.1 and 5.2.

Method 1: ignore inflation

Example 5.1 compared the cash flows from two proposals to purchase and operate a piece of equipment for five years. The comparison was made on the basis of present worth. The present worth of Proposal 1 was £14,366 and the present worth of Proposal 2 was £14,528. The cash flows leading to these present worths were based on estimates at year zero prices, the interest rate was taken at 15 per cent, and this comparison, which excluded inflation, indicated that Proposal 1 was the most economic.

Inflation would increase the running costs of both proposals and so increase the present worths. It should be remembered that the present worth is the sum of money required to be invested now to generate the stated cash flows, given a certain interest rate. Thus, if the cash flows are increased, the amount required for investment to generate the cash flows must also increase, given that the interest rate remains the same.

If the purpose of the exercise is simply to *compare* the proposals and select the most economic, and the inflation rate assumed is small, the comparison will not be affected seriously by the inclusion of inflation. That is, if *small* inflation allowances are added into both proposals, the difference between them will not be affected enough to change the comparison. This may be sufficient for small inflation rates but is unlikely to be satisfactory for larger inflation rates.

Method 2: adjusting the cash flows

Example 5.5

The cash flows in Example 5.1 are listed in Table 5.2. These cash flows were all estimated at year zero prices. Any cash flow not occurring in year zero would be subject to inflation. If an inflation rate of 10 per cent per annum were assumed the cash flows would be adjusted as shown in Table 5.3. The calculated cashflows are presented in Table 5.4. Now that the cash flows have been adjusted for inflation, the present worth of both proposals can be calculated as before. Because the cash flows vary from year to year and are not uniform, the present worth factors for each year will have to be used, as in Example 5.2.

Table 5.2 Cash flows for Proposals 1 and 2 for Example 5.5

Year	*Proposal 1 cash flow (£)*	*Proposal 2 cash flow (£)*
0	8,500	9,500
1	1,750	1,500
2	1,750	1,500
3	1,750	1,500
4	1,750	1,500
5	1,750	1,500

Table 5.3 Adjusted cash flows for Example 5.5

Year	*Proposal 1: Original cash flow + inflation adjustment (£)*	*Proposal 2: Original cash flow + inflation adjustment (£)*
0	8,500	9,500
1	1,750 + 10%	1,500 + 10%
2	(1,750 + 10%) + 10%	(1,500 + 10%) + 10%
3	((1,750 + 10%) + 10%) + 10%	((1,500 + 10%) + 10%) + 10%
4	(((1,750 + 10%) + 10%) + 10%) +10%	(((1,500 + 10%) + 10%) + 10%) +10%
5	((((1,750 + 10%) + 10%) + 10%) + 10%) + 10%	((((1,500 + 10%) + 10%) + 10%) + 10%) + 10%

The present worths of both proposals are calculated as follows, using an interest rate of 15 per cent.

Proposal 1

The present worth for Proposal 1 is calculated as shown in Table 5.5. The present worth factors were taken from the tables in the Appendix.

Proposal 2

The present worth for Proposal 2, similarly calculated, is £16,076.58. This comparison reverses the choice indicated in Example 5.1 by indicating that at 10 per cent inflation Proposal 2 becomes the more economic. Proposal 2, having the smaller running costs, suffers less from inflation than Proposal 1. At smaller inflation rates the choice would still be Proposal 1, as before. However, with an inflation rate at 10 per cent Proposal 2 becomes the more economic. Originally, the present worth of £14,366 for Proposal 1 was sufficient to provide for the £8,500 purchase price and the £1,750 running costs each year, and the present worth of £14,528 for Proposal 2 was sufficient to provide for the £9,500 purchase price and the £1,500 running costs each year. However, 10 per cent inflation causes the present worth of Proposal 1 to become £16,172.69 and provides for the £8,500 purchase price, the £1,750 running costs each year and the additional running costs incurred due to inflation.

Table 5.4 Calculated cash flows for Example 5.5

Year	*Proposal 1*					*Proposal 2*				
	Original cash flow (£)	+	*Inflation adjustment* (£)	=	*Adjusted cash flow* (£)	*Original cash flow* (£)	+	*Inflation adjustment* (£)	=	*Adjusted cash flow* (£)
0	8,500		0		8,500.00	9,500		0		9,500.00
1	1,750		175.00		1,925.00	1,500		150.00		1,650.00
2	1,750		367.50		2,117.50	1,500		315.00		1,815.00
3	1,750		579.25		2,329.25	1,500		496.50		1,996.50
4	1,750		812.18		2,562.18	1,500		696.15		2,196.15
5	1,750		1,068.39		2,818.39	1,500		915.77		2,415.77

Table 5.5 Proposal I present worth for Example 5.5

Year	*Cash flows (£)*	×	*Present worth factors (15%)*	=	*Present worth (£)*
0	8,500.00		1.0		8,500.00
1	1,925.00		0.86956		1,673.90
2	2,117.50		0.75614		1,601.13
3	2,329.25		0.65751		1,531.51
4	2,562.18		0.57175		1,464.93
5	2,818.39		0.49717		1,401.22
			Total present worth		16,172.69

Table 5.6 Proposal I adjusted cash flows for Example 5.6

Year	*Proposal I*	
	Original cash flows (£)	*Adjusted cash flows (£)*
0	8,500	8,500.00
1	1,750	1,925.00
2	1,750	2,117.50
3	1,750	2,329.25
4	1,750	2,562.18
5	1,750	2,818.39

The £16,076.58 present worth of Proposal 2 provides the £9,500 purchase price, the £1,500 running costs each year and smaller additional running costs due to inflation.

Method 3: adjusting the interest rate

The previous comparison including inflation could have been achieved by adjusting the interest rate rather than the cash flow.

Example 5.6

In Example 5.5 the cash flows from Example 5.1 were adjusted to include an inflation rate of 10 per cent per annum (refer to Table 5.6).

This adjustment was achieved by adding 10 per cent to the first year cash flows and 10 per cent + 10 per cent to the second year cash flows, and so on. This can be represented as shown in Table 5.7, where d represents the inflation rate (0.1 for 10 per cent).

To calculate the present worth of each of these adjusted cash flows, multiply each year by the present worth factor, as, for example, year three:

Year	Cash flow		Present worth factor
3	$£1,750 \times (1+d)^3$	×	0.65751

The present worth factor, 0.65751, was taken from the tables in the Appendix or calculated from the expression $1/(1+i)^n$, where i is the interest rate and n is the number of years. In this case $i = 0.15$ (for 15 per cent) and $n = 3$.

Table 5.7 Proposal I inflation adjustment for Example 5.6

Year	*Proposal I*		
	Original cash flow (£)	×	*Inflation adjustment*
0	8,500		
1	1,750		$(1+d)^1$
2	1,750		$(1+d)^2$
3	1,750		$(1+d)^3$
4	1,750		$(1+d)^4$
5	1,750		$(1+d)^5$

The present worth for year three can be calculated thus:

Year	Cash flow		Present worth factor
3	$£1{,}750.00 \times (1+d)^3$	×	$1/(1+0.15)^3$

or for any year as:

Year	Cash flow		Present worth factor
n	$£1{,}750.00 \times (1+d)^n$	×	$1/(1+i)^n$

This calculation can be simplified by the following adjustment: for $(1+i)^n$, substitute $(1+d)^n$ $(1+e)^n$, where d is the inflation rate as before and e is calculated such that:

$$(1+i)^n = (1+d)^n\,(1+e)^n$$

giving:

$$(1+e) = \frac{(1+i)}{(1+d)}$$

and

$$e = \frac{(1+i)}{(1+d)} - 1$$

Using this substitution, the calculation of present worth becomes:

Year	Cash flow		Present worth factor
n	$£1{,}750 \times (1+d)^n$	×	$\frac{1}{(1+d)^n(1+e)^n}$

The elements $(1+d)^n$ cancel and the calculation is reduced to:

$$£1{,}750 \times \frac{1}{(1+e)^n}$$

The present worth in the original example was calculated by multiplying the cash flow by the present worth factor $1/((1+i)^n)$, and so the present worth calculated above differs only in the interest rate used. The inflation adjustment has been transferred from the cash flow to the interest rate.

Given the interest rate $i = 0.15$ (15 per cent) and the inflation rate $d = 0.10$ (10 per cent):

$$e = ((1 + 0.15)/(1 + 0.10)) - 1 = 0.0454545 = 4.54\%$$

Therefore, the present worth of Proposal 1, using the interest rate adjusted for inflation is as shown in Table 5.8. The present worth calculated from the original cash flows and the adjusted interest rate gives a value of £16,172.69, which is the same as that given by Example 5.5, where the interest rate was kept at 15 per cent and the cash flows were adjusted for inflation.

The adjusted interest rate of 4.54 per cent is measuring the interest earned in excess of the inflation rate, and by taking the effect of inflation away from the interest rate it is possible to calculate present worths which allow for the effect of inflation. Calculating the present worth with the adjusted interest rate involves less work than first adjusting the cash flows and then calculating the present worth. Consequently, the method of adjusting the interest rate to allow for inflation is the most commonly employed. This technique allows for the effects of inflation by reducing the actual interest earned from the apparent rate to an 'effective' or 'real' rate.

In Example 5.6 the interest rate was 15 per cent. This would be the rate that the investor would take to represent the value of money and in all probability would be equated to an interest rate that could be earned in investments deposited elsewhere. The inflation rate used was 10 per cent and the 'effective' interest was calculated at 4.54 per cent for the following reasons. £100 at present-day prices would, at the end of the first year, with inflation at 10 per cent, be equivalent to £110, and the amount required for investment today at 15 per cent to produce £110 in one year would be £110 × 0.86956 = £95.65 (0.86956 is the present worth factor for 15 per cent). If the £100 required at the end of one year were left and the effect of inflation subtracted from the interest rate, then the amount required to be invested today at 4.54 per cent would be £100 × 0.95652 = £95.65, where 0.95652 is the present worth factor for 4.54 per cent. Thus, the amount required for investment today would be the same.

Table 5.8 Proposal 1 present worth for Example 5.6

Year	*Proposal 1*		
	Cash flow (£)	*Present worth factors*	*Present worth (£)*
0	8,500	1	8,500.00
1	1,750	0.95652	1,673.90
2	1,750	0.91493	1,601.13
3	1,750	0.87515	1,531.51
4	1,750	0.83710	1,464.93
5	1,750	0.80070	1,401.22
		Total present worth	16,172.69

Another example would be to assume an interest rate of 15 per cent and an inflation rate of 15 per cent. Since the inflation rate and the rate at which interest is earned are the same, the effective rate becomes zero, as follows:

$$e = ((1 + i)/(1 + d)) - 1 = ((1.15)/(1.15)) - 1 = 0$$

Thus, if £100 at present day prices were due at the end of one year and inflation at 15 per cent made this £115, the amount required to be invested today at 15 per cent to produce £115 in one year would be £115 × 0.86956 = £100 (0.86956 is the present worth factor for 15 per cent). If the £100 required at the end of one year were left and the effect of inflation taken away from the interest rate, then the amount required today at zero per cent would be £100 × 1.0 = £100, where 1.0 is the present worth factor for zero per cent. The amount required by both calculations is the same.

This example is particularly noteworthy because, if inflation rates and interest rates are equal, calculating present worths taking account of inflation simply involves summing the cash flows estimated at present prices. The effect of inflation totally eliminates the earned interest.

A final example would be to assume an interest rate of 15 per cent, as before, but an inflation rate of 20 per cent so that the inflation rate is greater than the rate at which interest can be earned. The effective rate then becomes −4.16 per cent as follows:

$$e = ((1 + i)/(1 + d)) - 1 = ((1.15)/(1.20)) - 1 = -0.0416$$
$$= -4.16\%$$

Thus, if £100 at present-day prices were due at the end of one year and inflation at 20 per cent would make this £120, the amount required to be invested today at 15 per cent to produce £120 in one year would be £120 × 0.86956 = £104.34, where 0.86956 is the present worth factor for 15 per cent. If the £100 required at the end of one year were left and the effect of inflation taken away from the interest rate, then the amount required today at −4.16 per cent would be £100 × $1/(1 - 0.0416)^1$ = £100 × 1.0434 = £104.34, where 1.0434 is the present worth factor for −4.16 per cent. It is to be noted that the present worth factor had to be calculated from the expression $1/(1 + i)^n$ because negative interest rates are not usually tabulated. Again the amounts required calculated by the two methods are the same.

Thus, this method of adjusting the interest rate is valid for *all* interest and inflation rates and, by producing cash flows estimated on the basis of present-day prices, the effect of inflation at various rates can be easily assessed, using a range of assumed inflation rates. The following example illustrates the application of this technique to Example 5.2.

Example 5.7

Year	Cash flow for the purchase, operating and resale of an item of equipment (£)
0	−8,500
1	−1,750
2	−1,850
3	−2,000
4	−2,200
5	−2,500 + 4,000

Table 5.9 Proposal 1 present worth allowing for 12% inflation for Example 5.7

Year	*Cash flow (£)*	*Present worth factors*	*Present worth (£)*
0	−8,500	1.0	−8,500.00
1	−1,750	0.9739	−1,704.32
2	−1,850	0.9485	−1,754.73
3	−2,000	0.9237	−1,847.40
4	−2,200	0.8996	−1,979.12
5	−2,500	0.8761	−2,190.25
5	+4,000	0.8761	+3,504.40
		Total present worth	−14,471.42

The cash flows estimated at present-day prices reflect only the increasing cost of operating the equipment and not increases due to inflation.

The value of money is 15 per cent. The present worth, as shown in Example 5.2, is £13,247.72. If inflation over the next five years is estimated at 12 per cent, the effective interest rate would be 2.68 per cent, calculated as follows

$$e = ((1 + i)/(1 + d)) - 1 = 1.15/1.12 - 1 = 0.0268 = 2.68\%$$

The present worth of the cash flows, allowing for inflation at 12 per cent, is £14,471.42, calculated as presented in Table 5.9. This £14,471.42 is the present worth of the original cash flows plus the present worth of the additional cash flows that would have to be included for inflation.

Method 4: varying inflation rates

Methods 2 and 3 illustrate how uniform inflation rates can be dealt with, but this leaves the difficulty of coping with varying inflation rates. The most commonly adopted approach in these economic comparisons is to take a long-term view of the interest rate to be used to represent the value of money and to ignore short-term variations.

The same argument is usually applied to the assumed inflation rates. It is possible, if required, to cope with varying inflation rates but to do this by adjusting the cash flows as illustrated in Method 3 for uniform inflation rates. The following example based on Example 5.5 demonstrates how cash flows can be adjusted for varying inflation rates.

The assumed inflation rates are 10 per cent for years one and two, 12 per cent for year three and 14 per cent for years four and five.

Example 5.8

Proposal 1

Table 5.10 shows the cash flow adjustments for Proposal 1. When calculated, these figures become:

Table 5.10 Varying inflation rates for Example 5.8

Year	*Original cash flows (£)*	*Proposal 1: inflation adjustments*				
		For year 1	*For year 2*	*For year 3*	*For year 4*	*For year 5*
0	8,500.00					
1	(1,750.00	+10 %)				
2	((1,750.00	+10 %)	+10 %)			
3	(((1,750.00	+10 %)	+10 %)	+12 %)		
4	((((1,750.00	+10 %)	+10 %)	+12 %)	+14 %)	
5	(((((1,750.00	+10 %)	+10 %)	+12 %)	+14 %)	+14 %)

Year	Proposal 1 cash flows including inflation adjustment (£)
0	8,500.00
1	1,925.00
2	2,117.50
3	2,371.60
4	2,703.62
5	3,082.13

Similarly, the cash flows for Proposal 2 can be adjusted for inflation as follows:

Year	Proposal 2 cash flows including inflation adjustment (£)
0	9,500.00
1	1,650.00
2	1,815.00
3	2,032.80
4	2,317.39
5	2,641.83

The present worths of both proposals, calculated at 15 per cent, are £16,412.50 for Proposal 1 and £16,282.14 for Proposal 2. Thus, with this inflation pattern and interest at 15 per cent, Proposal 2 is the more economic.

Note on all methods

All the examples shown based the comparison on present worth and did not employ EACs. The reason for this is that EAC comparisons involve more calculation than present worth when the annual cash flows are not uniform. When inflation is introduced, the annual cash flows cannot be uniform and it is easier to use present worth calculations.

Valuation of an item of equipment

The principles of economic comparisons can be employed to place a value on an item of equipment. The term 'value' has several definitions, such as the accountant's value as recorded in the asset register or the market value as determined by how much the item will sell for on the open market. The market value has the most practical meaning, as it is the amount of capital that can be obtained for the equipment item. However, the equipment item may be worth more to the owner than he can obtain by selling the equipment or the market value may be more than the equipment is worth to the owner. To determine either of these values requires the evaluation of worth or value to the owner of the item of equipment. Economic comparisons, based on present worth, can be used to determine the value of the equipment to the owner.

Example 5.9

An item of equipment, the original purchase price of which was £10,000, and which has been in use for two years, has a remaining useful life of four years and the estimated running costs at present prices for the next four years are £661.25, £766.44, £874.50 and £1,005.68. The estimated resale value at the end of the four years is £2,000. Thus, the cash flows are:

Year	Cash flows for existing plant item (£)
0	
1	−661.25
2	−766.44
3	−874.50
4	−1,005.68 + 2,000

Note that there is no cash flow in year zero, because the plant item is already owned. The present worth of these cash flows, using an interest rate of 10 per cent, is £1,212.45, calculated as shown in Table 5.11.

If this item of equipment were not available to the owner, an alternative method of providing the equipment would be necessary and the present worth of the alternative method would have to be calculated and compared with that for the equipment already owned.

Alternative method 1: hiring

If the annual hire charge for a similar item of equipment were £2,500 each year for four years, then the present worth of hiring this equipment for four years, using an interest rate

Table 5.11 Present worth calculations for Example 5.9

Year	*Cash flows for existing equipment item (£)*	*Present worth factors (10%)*	*Present worth (£)*
0	–	–	–
1	−661.25	0.90909	−601.14
2	−766.44	0.82644	−633.42
3	−874.50	0.75131	−657.02
4	−1,005.68 + 2,000	0.68301	+679.13
		Total present worth	−1,212.45

of 10 per cent, would be £7,924.50, calculated as:

$$£2{,}500 \times 3.1698 = £7{,}924.50$$

The difference between the two present worths is an estimate of the value of the equipment to the owner. The difference is £7,924.50 − £1,212.45 = £6,712.05, and if the owner were able to sell the equipment for £6,712.05 and to offset this against the cost of hiring, then the present worth of cash flows for hiring would be exactly the same as that for already owning the equipment. If the owner were able to sell the equipment for more than £6,712.05, it would be more economic to do so and to hire the equipment. If £6,712.05 could not be realised by selling the existing equipment, it would be more economic to retain the existing item of equipment.

If the effects of inflation were to be introduced, this could be achieved by adjusting the interest rates as explained previously, in the section dealing with inflation adjustments.

Alternative method 2: buying new equipment

If a similar item of equipment were available for purchase, then the cash flows for purchase, operating and resale would have to be estimated as in the next example.

Example 5.10

The cash flows estimated at present prices for purchasing, operating and reselling after four years for a similar item of equipment are:

Year	Cash flow for purchase, operating and resale for new item of equipment (£)
0	−11,000.00
1	−400.00
2	−460.00
3	−529.00
4	−608.35 + 4,000

The present worth of these cash flows, using an interest rate of 10 per cent, is calculated as shown in Table 5.12.

The difference between the present worth of keeping the existing equipment and replacing immediately with a new item of equipment is: £9,824.71 − £1,212.45 = £8,612.26. Thus, if the owner could sell the existing equipment for £8,612.26 and offset this against the cost of the new equipment, the present worth of acquiring the new equipment would be the same as keeping the existing item of equipment. If the owner could sell the item of equipment for more than £8,612.26, then it would be more economic to sell and replace; if £8,612.26 could not be realised from the sale of the equipment, it would be more economic to retain the equipment.

If the effects of inflation were to be introduced, this could be achieved by adjusting the interest rates as explained previously, in the section dealing with inflation adjustments.

Table 5.12 Present worth calculation for buying new equipment for Example 5.10

Year	*Cash flows for new item of equipment (£)*	*Present worth factors (10%)*	*Present worth (£)*
0	−11,000.00	1.0	−11,000.00
1	−400.00	0.90909	−363.64
2	−460.00	0.82644	−380.16
3	−529.00	0.75131	−397.44
4	−608.35 + 4,000	0.68301	+2,316.53
		Total present worth	−9,824.71

Valuations for longer lives

The above valuations have been calculated from the useful life of the existing item of equipment. This was estimated at four years, the equipment already being two years old and having a total life of six years. If the need for the equipment extended for, say, 20 years, the comparison would also need to be extended for that time period. Thus, the cost of keeping the existing equipment and its subsequent replacements must be compared with the cost of immediate replacement and subsequent replacements. The replacements will all be estimated at the same costs as the immediate replacements, as all cash flows are estimated at present prices and the effects of inflation incorporated separately.

Example 5.11

The cash flows for keeping the existing item of equipment and subsequent replacements for 20 years are as illustrated in Table 5.13 whilst the cash flows for immediate replacement with a new item of equipment and replacements for 20 years are provided in Table 5.14.

To compare these cash flows, the present worth of both must be calculated and the present worth of the immediate replacement can be calculated as shown in Table 5.15, using an interest rate of 10 per cent.

Thus, the cash flows for the immediate and subsequent replacements can be represented as:

Year	Cash flow (£)
0	−12,316.35
6	−12,316.35
12	−12,316.35
18	−11,000.00
19	−400.00
20	−460.00 + 8,000

The £12,316.35 at years zero, six and 12 represents all the cash flows for the replacements purchased in those years, and the present worth for the immediate and subsequent replacements up to 20 years is £24,115.92, calculated as shown in Table 5.16.

Table 5.13 Cash flows for retention/subsequent replacements for Example 5.11

Year	Cash flows				
	Running costs of existing equipment (£)	*Resale of existing equipment (£)*	*Purchase of replacements (£)*	*Running costs of replacements (£)*	*Resale of replacements (£)*
0	–				
1	−661.25				
2	−766.44				
3	−874.50				
4	−1,005.68	+2,000.00	−11,000.00		
5				−400.00	
6				−460.00	
7				−529.00	
8				−608.35	
9				−699.60	
10			−11,000.00	−804.54	+2,000
11				−400.00	
12				−460.00	
13				−529.00	
14				−608.35	
15				−699.60	
16			−11,000.00	−804.54	+2,000
17				−400.00	
18				−460.00	
19				−529.00	
20				−608.35	+4,000

Table 5.14 Cash flows for immediate/subsequent replacement for Example 5.11

Year	Cash flows		
	Purchase price (£)	*Running costs (£)*	*Resale (£)*
0	−11,000		
1		−400.00	
2		−460.00	
3		−529.00	
4		−608.35	
5		−699.60	
6	−11,000	−804.54	+2,000
7		−400.00	
8		−460.00	
9		−529.00	
10		−608.35	
11		−699.60	
12	−11,000	−804.54	+2,000
13		−400.00	
14		−460.00	
15		−529.00	
16		−608.35	
17		−699.60	
18	−11,000	−804.54	+2,000
19		−400.00	
20		−460.00	+8,000

Table 5.15 Present worth of immediate/subsequent replacement for Example 5.11

Year	*Cash flow (£)*	*Present worth factor (10%)*	*Present worth (£)*
0	−11,000.00	1.0	−11,000.00
1	−400.00	0.90909	−363.64
2	−460.00	0.82644	−380.16
3	−529.00	0.75131	−397.44
4	−608.35	0.68301	−415.51
5	−699.60	0.62092	−434.40
6	−804.54 + £2,000.00	0.56447	+674.80
		Total present worth	−12,316.35

Table 5.16 Present worth of immediate/subsequent replacements up to 20 years for Example 5.11

Year	*Cash flow (£)*	*Present worth factor (10%)*	*Present worth (£)*
0	−12,316.35	1.0	−12,316.35
6	−12,316.35	0.56447	−6,952.21
12	−12,316.35	0.31863	−3,924.36
18	−11,000.00	0.17985	−1,978.35
19	−400.00	0.16350	−65.40
20	−460.00 + £8,000	0.14864	+1,120.75
		Total present worth	−24,115.92

The present worth of keeping the existing equipment until the end of its useful life and its subsequent replacement can be calculated first by representing the cash flows as shown:

Year	Cash flows (£)
0	−1,212.45
4	−12,316.35
10	−12,316.35
16	−11,000.00
17	−400.00
18	−460.00
19	−529.00
20	−608.35 + 4,000

The £1,212.45, taken from Example 5.9, represents all the cash flows for the existing item of equipment.

The £12,316.35 in years four and ten is used to represent all the cash flows for the replacement in those years. Consequently, the present worth for keeping the existing plant item until the end of its useful life and subsequent replacements up to 20 years is £16,511.14, calculated as shown in Table 5.17.

The difference between these two present worths is £24,115.92 − £16,511.14 = £7,604.78. If the owner could sell the existing equipment for £7,604.78 and this amount could be offset

Table 5.17 Present worth of retention/subsequent replacements for Example 5.11

Year	*Cash flows (£)*	*Present worth factor (10%)*	*Present worth (£)*
0	−1,212.45	1.0	−1,212.45
4	−12,316.35	0.68301	−8,412.19
10	−12,316.35	0.38554	−4,748.45
16	−11,000.00	0.21762	−2,393.82
17	−400.00	0.19784	−79.14
18	−460.00	0.17985	−82.73
19	−529.00	0.16350	−86.49
20	−608.35 + £4,000	0.14864	+504.13
		Total present worth	−16,511.14

against the cost of the immediate replacement, the present worth of the immediate and subsequent replacements would be the same as keeping the existing item. If the owner could sell the existing equipment for more than £7,604.78, the acquiring of an immediate replacement would be more economic. If £7,604.78 could not be realised by the sale of the existing equipment, then keeping the existing equipment would be more economic.

If the effects of inflation were to be introduced, this could be achieved by adjusting the interest rates as explained previously, in the section dealing with inflation adjustments.

The above calculation was based on an overall duration of 20 years, and the present worths were comparable because they were calculated for equal periods of time. The life of the equipment is six years. With immediate and subsequent replacements the 20 years is made up by disposing of the replacement in year 18 after two years, and by adjusting the resale value of this last replacement. Similarly, the last replacement in the case of keeping the existing equipment and replacing from year four onwards occurs in year 16, and again the resale value of this last replacement has been adjusted.

Example 5.12

An alternative to assuming a finite cut-off of 20 years, as in the last example, is to assume that the replacements will continue in perpetuity. The cash flow for immediate and subsequent replacements to infinity can be represented as follows:

Year	Cash flows representing replacements to infinity (£)
0	−12,316.35
6	−12,316.35
12	−12,316.35
18	−12,316.35
24	−12,316.35
...	...
...	...
...	...
∞	∞

The £12,316.35 is the present worth of purchasing, operating and reselling the new equipment. This present worth was calculated previously as part of Example 5.11. If this equipment is replaced every six years, this amount recurs every six years to infinity.

The present worth of a sum recurring every year, starting at the end of the first year, is given by the factor $((1+i)^n - 1)/(i(1+i)^n)$ which is the uniform series present worth factor given in the Appendix.

The present worth of a sum recurring not every year but every y years, starting in y years, is given by the factor $1/((1+i)^y - 1)$. Thus, if the sum recurring every y years were x, the present worth would be

$$x \times \left(\frac{1}{(1+i)^y - 1}\right)$$

and if the sum x occurred in year zero, the total present worth of the whole series would be

$$x + x\left(\frac{1}{(1+i)^y - 1}\right)$$

which reduces to

$$\frac{x}{1 - (1/(1+i))^y}$$

Thus, substituting £12,316.35 for x, six years for y and 10 per cent for i, the present worth of the series starting in year zero and recurring to infinity is

$$\frac{\text{£}12{,}316.35}{1 - 0.56447} = \text{£}28{,}278.99$$

The 0.56447 can be taken from tables, since the element $1/(1+i)^y$ is the expression for the present worth factor.

The amount calculated, £28,278.99 represents the amount that would be required today which, if invested at 10 per cent, would produce £12,316.35 every six years forever. This can be checked as follows. First take away the initial £12,316.35; this leaves £15,962.64 for investment. In six years this £15,962.64 would increase to £15,962.64 × 1.77156 = £28,278.99. This £28,278.99 is made up of the original capital £15,962.64, and interest earned in those six years of £12,316.35. The factor 1.77156 is the compound amount factor for 10 per cent and is taken from tables in the Appendix. Thus, the sum of £12,316.35 can be used every six years to provide replacement equipment. This £12,316.35 is the interest earned on the capital of £15,962.64 and is entirely used up every six years. The capital of £15,962.64 remains undepleted and can go on producing £12,316.35 every six years indefinitely.

The £12,316.35 is the present worth of purchasing, operating and reselling the equipment item, and £28,278.99 is the present worth of purchasing, operating and reselling every six years in perpetuity. This present worth is sometimes called the capitalised cost.

To arrive at a valuation of the existing equipment, it is necessary to calculate the capitalised cost of the cash flows relating to the scheme whereby the existing equipment is

Table 5.18 Cash flows for keeping existing equipment and its replacements for Example 5.12

Year	*Cash flows for keeping existing equipment and its replacements*		
	Running costs of existing equipment (£)	*Resale (£)*	*Replacements (£)*
0	–		
1	−661.25		
2	−766.44		
3	−874.50		
4	−1,005.68	+2,000	
10			−12,316.35
16			−12,316.35
22			−12,316.35
28			−12,316.35
...			...
...			...
...			...
∞			∞

retained – that is, the present worth of the existing equipment and its replacements to infinity. Taking the cash flows from Example 5.11 for keeping the existing equipment and its replacements these can be represented as shown in Table 5.18.

The cash flows in years 1–4 representing the running costs of the existing equipment and the resale value have a present worth of £1,212.45, as calculated in Example 5.9.

The £12,316.35 every six years has a present worth of £28,278.99, as calculated in Example 5.11. However, in this set of cash flows the first of the £12,316.35 occurs in year four and the cash flows for keeping the existing equipment and its replacements to infinity can be represented as follows:

Year	Cash flows for keeping the existing equipment and its replacements (£)
0	−1,212.45
4	−28,278.99

The present worth of these cash flows is £20,527.28, calculated thus:

$$£1{,}212.45 + (£28{,}278.99 \times 0.68301) = £20{,}527.28$$

Given an interest rate of 10 per cent, £20,527.28 is the amount required for investment today to give £1,212.45 in year zero, enough for the existing equipment, and £28,278.99 in year four, enough for the replacements at £12,316.35 every six years until infinity. Thus, £20,527.28 is the capitalised cost of keeping the existing equipment and then replacing it in perpetuity.

The difference between the capitalised cost of keeping the existing equipment and its immediate replacement is the value to the owner of the existing equipment. The difference is £28,278.99 – £20,527.28 = £7,751.71. This £7,751.71 is the amount that could be compared with the market value when considering disposal. The value calculated here is only slightly different from that calculated for replacements up to 20 years, because the present worth factors become smaller with increasing time and the effect of cash flows beyond 20 years on such calculations is small.

If the effects of inflation were to be incorporated in this calculation, this would be achieved by adjusting the interest rates as explained previously, in the section dealing with inflation adjustments.

Determining replacement age

The factors that determine the economic replacement age of equipment are the purchase price, the operating costs and the resale value. The purchase price is relevant because the equipment must be kept long enough to warrant the investment. The operating costs usually increase with ageing equipment and it is, therefore, important not to keep the equipment too long. Also, as these operating costs increase the resale value declines. The purpose of an economic analysis is to find the balance between these. As the comparisons are being made between keeping equipment for one, two, three or four years, etc., the use of present worth presents the difficulty of equalising the lives of the comparisons. This may be achieved by considering replacements to infinity. As the operating costs are not usually uniform, the use of EACs is not easily applied either. Thus, neither present worth nor EACs offer any real advantage in this comparison. The method presented here is based on EACs.

Determining replacement age using equivalent annual costs

The concept of EACs is explained in Example 5.3.

Example 5.13

The purchase price of a small concrete batching unit is £25,000. The operating costs, based on the estimated annual average hours of use, are £1,000 in the first year when manufacturers' warranties operate, and £1,500 in the second year, rising by £375.00 per year thereafter. The resale values are:

Year	Predicted resale values (£)
1	22,500
2	20,000
3	18,750
4	15,000
5	10,000
6	6,250

The calculations to determine the EAC of keeping this equipment for one, two, three and four years, etc., are set out in Table 5.19. Most of the entries in Table 5.19, which calculate the EACs for one, two, three, etc., years, are self-explanatory.

Column *C* shows the EAC of the purchase price. For example, if the equipment were kept for four years, the EAC of the purchase price would be £8,750 per year.

Table 5.19 Determining replacement age by use of EACs

Year	A	B	C	D	E	F	G	H	I	K	L	M	N
	Purchase price (£)	*Capital recovery factors for 15%*	*EAC of purchase price (A × B) (£)*	*Running costs (£)*	*Present worth factors at 15%*	*Present worth of running costs (D × E) (£)*	*Sum of present worth of running costs (ΣF) (£)*	*EAC of present worth of running costs (B × G) (£)*	*EACs of purchase price and running costs (C + H) (£)*	*Resale value (£)*	*Present worth of resale (K × E) (£)*	*EACs of resale (B × L) (£)*	*EACs of purchases running and resales (I − M) (£)*
0	25,000				1.0								
1		1.150	28,750	1,000	0.869	869.00	869.00	999.35	29,749.35	22,500	19,552.50	22,500.00	7,249.35
2		0.615	15,375	1,500	0.756	1,134.00	2,003.00	1,231.84	16,606.84	20,000	15,120.00	9,298.00	7,308.04
3		0.437	10,295	1,875	0.657	1,231.88	3,234.88	1,413.64	12,338.64	18,750	12,318.75	5,383.29	6,955.35
4		0.350	8,750	2,250	0.571	1,284.75	4,519.63	1,581.87	10,331.87	15,000	8,565.00	2,997.75	7,334.12
5		0.298	7,450	2,625	0.497	1,304.63	5,824.26	1,735.62	9,185.62	10,000	4,970.00	1,481.06	7,704.56
6		0.264	6,600	3,000	0.432	1,296.00	7,120.26	1,879.74	8,479.74	6,250	2,700.00	712.80	7,766.94

Note
The minimum in column N, £6,955.35, is the minimum EAC.

Column *F* is the present worth of the running costs. In year four the running costs for that year were £2,250, and the present worth of this amount is £1,284.75: that is, the amount required to be invested in year zero to provide £2,250 in year four is £1,284.75.

Column *G* is the running total of the present worths of the running costs, so £3,234.88 is the sum of the present worths for years one, two and three. Thus, if £3,234.88 were invested in year zero at 15 per cent it would provide enough to pay for the running costs of years one, two and three.

Column *H* is the EAC of the present worth of the running costs. Taking the £3,234.88 for year 3 from column *G* and converting it to an annual sum, using the capital recovery factor gives £1,413.64, and the running costs of £1,000, £1,500 and £1,875 for years one, two and three are now converted to a uniform series of £1,413.64 per year. The calculations in columns *F*, *G* and *H* are devices to convert the varying operating costs to a uniform series.

Column *I* is the EAC for the purchase price added to the EAC for the running costs.

Column *L* is the present worth of the resale value. £12,318.75 in year three is the present worth of the resale value of £18,750 in year three.

Column *M* is the EAC of the resale value. Thus, £5,383.28 for year three is the EAC of £12,318.75 from column *L*.

Column *N* is the EAC of the resale subtracted from the EACs of the purchase price and running cost. This gives the net EAC for purchase, operating and resale for years 1–6. Thus, the minimum in column *N* is the minimum EAC and the best replacement age by this economic criterion. The minimum is £6,955.35 in year three.

The effects of inflation could be incorporated by adjusting the interest rates that were used to calculate the present worth, but not the interest rate used to calculate the EACs.

Chapter 6

Equipment profitability and acquisition

Measuring profitability

The economic analyses described in Chapter 5 have related to schemes whereby only the expenditure was considered. The purchase price and running or operating costs were not offset in the calculations against any revenue, and the only monies returning to the investor were from resale. Thus, the previous analyses were confined to determining whether one course of action – say the purchase of one particular item of equipment – was more economic than another course of action, the purchase of an alternative item. What will be considered now is the situation where the equipment generates a revenue. The simplest situation to imagine is where the equipment is purchased and hired out so that the owner has capital expenditure, operating costs, revenue and resale value. The analysis required now is not simply whether one item is more economic than another but whether the item is earning an adequate return on the invested capital. That is, whether the return on capital derived from owning and hiring out equipment is better than could be obtained from less risky investment elsewhere or, if the capital to buy the equipment is borrowed, whether the return is greater than the cost of capital as measured by the interest on the capital. The capital made available by the organisation to purchase the equipment should earn at least the minimum return expected by the organisation. This analysis, therefore, requires that the rate of return be measured and the most widely used method is known as the 'internal rate of return', 'yield' or 'discounted cash flow (DCF) yield'. All these are names for the same measure of profitability.

The interest factors used in this chapter are taken from the Appendix.

Discounted cash flow yield

Calculating the DCF yield requires first that the net cash flows be calculated. The net cash flows are the sum of the cash flows relating to the investment project. It is useful in calculating the net cash flows to construct a cash flow tree or model as shown in Figure 6.1, a cash flow tree for a simple purchase, operating, hire and resale model. This shows the cash flows broken down into three levels. The first level is the net cash flow. It is from this net cash flow that the DCF yield will be calculated. The next level represents the three general categories of 'capital cost', 'operating cost' and 'operating revenue'. These three general categories apply to most cash flow trees.

It is the next level, level 3, that determines the unique features of the operation or investment being modelled, and breaks down the capital costs to purchase price and resale. The operating costs are broken down into estimates for 'maintenance', 'repairs', 'administration'

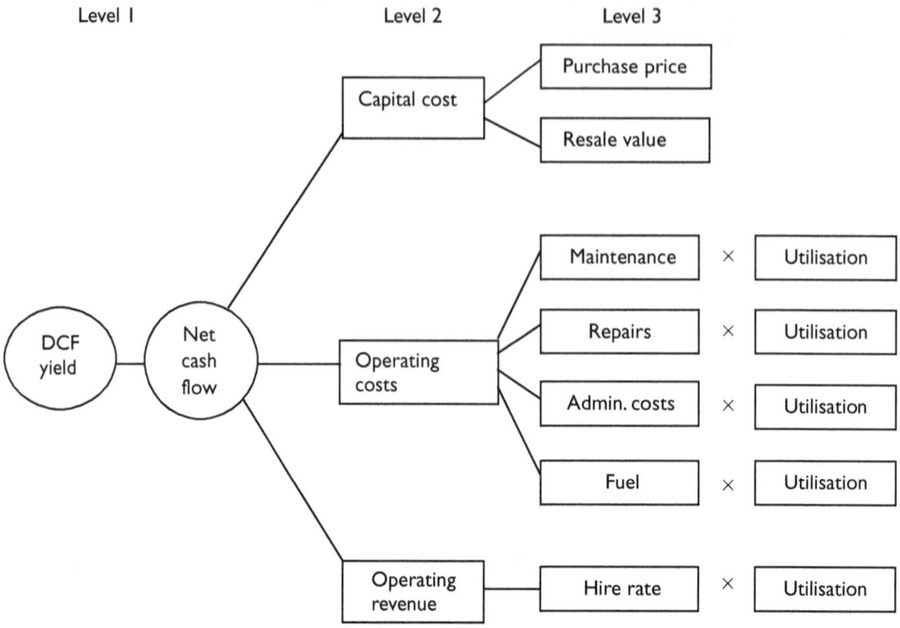

Figure 6.1 Simple cash flow tree for an equipment hire operation.

and 'fuel'. The estimate for the purchase price is a single sum occurring in year zero and the estimate for the resale value is a single sum occurring in the year when the resale takes place. The estimates for the operating costs are not single sums but estimates of cost for each year of operation. In more detailed models these estimates of cost may be made for each month. All the operating costs are shown tied to the utilisation, as clearly a relationship exists between the utilisation and these costs. Similarly, the operating revenue which is derived from the hire rate and the utilisation will be yearly or monthly estimates. If this level is not detailed enough for producing estimates, then each element in level 3 will have to be broken down further. When a level of detail is achieved such that estimates can be supplied, then the process of aggregation through the cash flow tree will produce the net cash flows, the algebraic sign being taken into account (positive cash flows being income and negative cash flows being outgoings). The DCF yield is calculated from these net cash flows.

To demonstrate calculating DCF yield and to explain the meaning and importance of DCF yield, an example will be used, based on the following simple net cash flows:

Year	Net cash flows (£)
0	−1,000.00
1	+315.47
2	+315.47
3	+315.47
4	+315.47

These cash flows are assumed to have been derived from a cash flow tree calculation similar to that described. These net cash flows show a pattern of negative, outgoing, cash flow in the beginning, followed by positive, incoming, cash flows in the subsequent years. The DCF yield or internal rate of return is a measure of the return on the capital invested of −£1,000.00, given by the positive cash flows occurring in years 1–4.

DCF yield or internal rate of return has two definitions:

1 The DCF yield or rate of return is the *maximum* interest rate that could be paid for borrowed capital, assuming that all the capital required to fund the project is acquired as an overdraft and all the positive cash flows are used to repay this overdraft.
2 The DCF yield or rate of return is the interest rate which, if used to discount a project's cash flow, will give a net present value (or worth) of zero.

The second of these definitions is more usually employed in calculating the yield, and is also the definition that has given rise to the name 'DCF yield'. It is also the more difficult to understand at first reading, so the example will be calculated in the first instance using the first definition.

Calculation of DCF yield by the first definition is a process of trial and error. An interest rate is assumed and tested to determine whether it is the maximum: if it is above the maximum, a new trial with a smaller interest rate is used; if it is below the maximum, a new trial with a larger interest rate is used until the maximum is found, eventually by interpolation, if necessary. Using an interest rate of 12 per cent as the first trial rate, the test as to whether this is the maximum is shown in Table 6.1.

In year zero the amount borrowed was −£1,000.00 and the interest on this during year one was −£120.00. Thus, at the end of year 1 the borrowing account was the original −£1,000.00 together with the interest of −£120.00 offset by the income of +£315.47, leaving −£804.53 in the borrowing account. Repeating this calculation until the end of the project indicates that there is −£65.79 left in the account. Thus, this project could not have paid interest at 12 per cent on the borrowed capital, because to do so would require £65.79 from other sources. Thus, 12 per cent is greater than the maximum interest rate that this project could support and the trial and error process continues with a smaller interest rate. Suppose that the guess is 8 per cent: the test as to whether this is the maximum is shown in Table 6.2.

Proceeding through the calculations as before, the amount left in the account at the end of the project is +£61.05. Thus, this project could have paid more for its borrowed capital than 8 per cent, which is therefore less than the maximum interest rate.

Given that 12 per cent is greater than the maximum interest rate and 8 per cent is less than the maximum interest rate, the maximum clearly lies between the two. Therefore, the next

Table 6.1 Calculation of DCF yield – trial one 12%

Year	*Net cash flow (£)*	*Interest paid on borrowed capital at 12% (£)*	*Borrowing account (£)*
0	−1,000.00	–	−1,000.00
1	+315.47	−120.00	−804.53
2	+315.47	−96.54	−585.60
3	+315.47	−70.27	−340.41
4	+315.47	−40.85	−65.79

Table 6.2 Calculation of DCF yield – trial two 8%

Year	*Net cash flow (£)*	*Interest paid on borrowed capital at 8% (£)*	*Borrowing account (£)*
0	−1,000.00	–	−1,000.00
1	+315.47	−80.00	−764.53
2	+315.47	−61.16	−510.22
3	+315.47	−40.82	−235.57
4	+315.47	−18.85	+61.05

Table 6.3 Calculation of DCF yield – trial three 10%

Year	*Net cash flow (£)*	*Interest paid on borrowed capital at 10% (£)*	*Borrowing account (£)*
0	−1,000.00	–	−1,000.00
1	+315.47	−100.00	−784.53
2	+315.47	−78.45	−547.51
3	+315.47	−54.75	−286.79
4	+315.47	−28.68	0.00

reasonable guess would be 10 per cent. The trial to determine whether 10 per cent is the maximum is illustrated in Table 6.3.

Performing the calculation as before shows that an amount of zero would be left in the borrowing account. Thus, 10 per cent is the maximum interest rate that could be paid for the borrowed capital in this project, and by definition (1) 10 per cent is the DCF yield.

Measuring this *maximum* that *could* be paid for the borrowed capital is a way of measuring how much the project cash flows are producing. Measuring the amount that can be taken away (by interest charges) shows how much the project is producing. Also, this measure (the DCF yield or rate of return) is directly comparable with the cost of capital, the cost of capital being the weighted average of the costs of all sources of capital used by the organisation. If the organisation were paying more for its capital than 10 per cent, the project would not be satisfactory, because it would not be yielding more than borrowing the capital is costing.

The method more commonly employed in calculating the DCF yield is according to definition (2). Calculating the DCF yield by the second definition is also a trial and error process which also requires an assumed interest rate and a trial to determine whether the assumed interest rate gives a net present value (NPV) or worth of zero. If the calculated NPV is negative the assumed interest rate is too large and a smaller one is assumed and the NPV is recalculated. If the calculated NPV is positive the assumed interest is too small. The process is repeated until the interest rate which gives a zero NPV is found – by interpolation, if necessary.

Using 9 per cent as a first trial, the NPV is calculated and found to be positive; a larger interest rate of 11 per cent is then used to produce a negative NPV (refer to Table 6.4).

Interpolation produces the interest rate which gives an NPV of zero and is found to be 10 per cent, as before.

Table 6.4 Calculating present worth based on uniform revenues

Year	*Net cash flow (£)*	*Present worth factors for 9% (1st trial)*	*Present worth (£)*	*Present worth factors for 11% (2nd trial)*	*Present worth (£)*
0	−1,000.00	1.0	−1,000.00	1.0	−1,000.00
1	+315.47	0.91743	+289.42	0.90090	+284.21
2	+315.47	0.84168	+265.52	0.81162	+256.04
3	+315.47	0.77218	+243.60	0.73119	+230.67
4	+315.47	0.70842	+223.49	0.65873	+207.81
			+22.03		−21.27

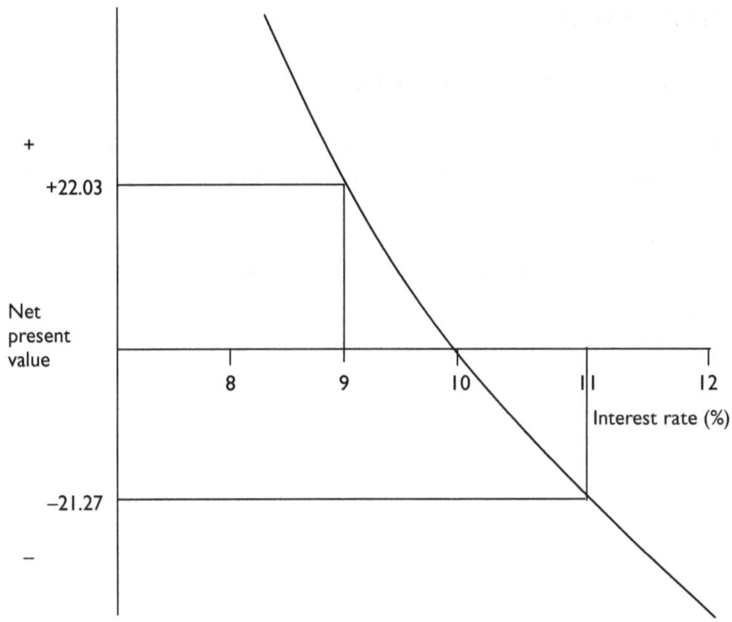

Figure 6.2 Graph of net present value against interest rate.

By interpolation: the interest rate which gives an NPV of zero

$$= 9\% + (11\% - 9\%) \times \left(\frac{22.03}{22.03 - (-21.27)}\right)$$

$$= 9\% + 2\% \times 0.508$$

$$= 9\% + 1.01\% = 10.0\%$$

The present worth factors are taken from tables or calculated as $1/(1 + i)^n$. This 10.0 per cent has exactly the same meaning as the 10.0 per cent calculated by the first definition. The trial and error process that produces the interest rate that gives an NPV of zero can be understood by examining the graph of NPV against interest rates as shown in Figure 6.2. Any minor discrepancy between the interpolated value and 10 per cent is due to interpolation which assumes that the graph in Figure 6.2 is linear between 9 and 11 per cent. It is, in fact, slightly curved.

To understand further the meaning of this calculated interest rate which has been given the name 'DCF yield' or 'internal rate of return' consider this question: if £1,000 were invested at 10 per cent, what regular income could be taken each year for the next four years?

The capital recovery factor for 10 per cent and four years is 0.31547. This factor is taken from the tables in the Appendix. Thus, the income that could be taken each year for four years from £1,000 invested at 10 per cent is:

$$£1{,}000 \times 0.31547 = £315.47$$

This £315.47 is the same as the net cash flow in the example used.

Given a capital sum and an interest rate, the income can be calculated. Also, given the capital sum and the income, the interest rate that would produce that income from that capital sum can be calculated. This interest rate is called the DCF yield or the internal rate of return.

This example was based on uniform revenues, but the searching techniques for DCF yield will work equally well for non-uniform revenues, as shown in the next example, in which the net cash flows are:

Year	Net cash flows (£)
0	−10,000
1	+2,800
2	+3,000
3	+2,500
4	+2,300
5	+2,200

The DCF yield is 9.35 per cent, calculated using two trial interest rates of 9 and 11 per cent and then interpolation (as shown in Table 6.5).

Table 6.5 Calculating present worth based on non-uniform revenues

Year	*Net cash flow (£)*	*Present worth factors for 9% (1st trial)*	*Present worth (£)*	*Present worth factors for 11% (2nd trial)*	*Present worth (£)*
0	−10,000	1.0	−10,000.00	1.0	−10,000.00
1	+2,800	0.91743	+2,568.80	0.90090	+2,522.52
2	+3,000	0.84168	+2,525.04	0.81162	+2,434.86
3	+2,500	0.77218	+1,930.45	0.73119	+1,827.97
4	+2,300	0.70842	+1,629.36	0.65873	+1,515.07
5	+2,200	0.64993	+1,429.84	0.59345	+1,305.59
		Net present worth	+83.50		−393.99

Interpolation:

$$\text{DCF yield} = 9\% + (11\% - 9\%) \times \left(\frac{83.50}{83.50 - (-393.99)}\right)$$

$$= 9\% + (2\% \times 0.175)$$

$$= 9\% + 0.35\%$$

$$= 9.35\%$$

The yield or rate of return is the most widely used measure of profitability.

Other measures of profitability

Yield is not the only measure of profitability: others in use are NPV, payback period and average annual rate of return.

Net present value is used to determine whether a proposed project yields at least the minimum return specified by the organisation. The NPV is calculated for the net cash flows, using the minimum rate of return required, this rate representing the cost of capital to the organisation. If the NPV is positive, it follows that the yield is above the minimum and the project is worthy of further consideration. If the NPV is negative, the yield is less than the minimum and the project can be rejected without further analysis.

The *payback period* is the time taken to repay the original capital invested. The payback period is not very useful without predetermining what a satisfactory payback period should be. If an organisation usually expects payback periods of one year or eight months it would clearly be unhappy with proposals that incurred payback periods of three or four years. In general, the shorter the payback the more profitable the project, provided that the project continues to have positive net revenues for a number of years after the payback period.

The *average annual rate of return* is all the returns (positive cash flows) for the project averaged over the number of years the project lasts and expressed as a percentage of the invested capital. Like the payback period, the average annual rate of return is not very useful without first determining what a satisfactory rate should be. An average annual rate of return of 33 per cent and a payback period of three years are similar. The higher this rate of return the more profitable the project.

The payback period or the average annual rate of return are never used on their own as measures of profitability but always in conjunction with other measures, such as NPV or even DCF yield.

Varying hire rates and yield

If a cash flow tree such as that illustrated in Figure 6.1 is constructed to represent an equipment hire operation, then the effects of the key elements in that cash flow tree can be studied by substituting a range of values for the key variables and determining the different net cash flows and, hence, the different yields. Two major variables in an equipment hire operation are the hire rate and the utilisation factor. So, if a range of five hire rates is used with a range of seven utilisation factors from 50 to 110 per cent, different net cash flows can be

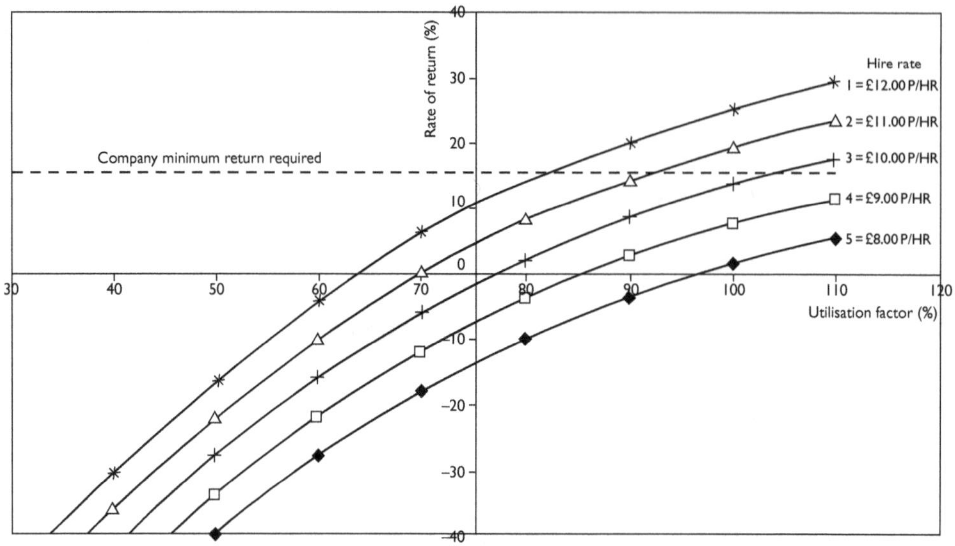

Figure 6.3 Sensitivity of the rate of return to the hire rate and the utilisation of an item of equipment.

calculated for each of the combinations, making a total of 35. For each of the net cash flows, a yield can be calculated. Results of this type are best presented graphically, as shown in Figure 6.3. This graph is known as a sensitivity chart and the analysis performed is known as a sensitivity analysis, as it is displaying the sensitivity of yield to both the hire rate and the utilisation factor.

This family of curves illustrates two obvious points: the greater the hire rate, and the greater the utilisation factor, the greater the rate of return. But while these points are obvious, this type of graph quantifies the increase in rate of return for any assumed increase in hire rate or utilisation factor. For example, for hire rate number three, £10 per hour, the graph illustrates that the plant has to be on hire for 77 per cent of the normal maximum usage before it would show a positive return, and to record a return equal to the organisation's minimum at this hire rate, the equipment would need to be on hire all the available normal working time plus some overtime working. Adding the organisation's minimum return required to the graph reveals the combination of usage rate and hire rate that would produce this return or more. The hire rate eventually adopted would be chosen for marketing reasons. These sensitivity graphs help to assess whether the market hire rate is likely to produce an adequate return on capital.

Yield and the effect of corporation tax and capital allowances

If an organisation trades profitably, then it will be subject to corporation tax. Currently, the current rate is 30 per cent. There is a delay between the organisation engaging in trade, declaring a profit and paying tax. This lag in paying tax varies and depends on the relationship between the organisation's accounting year and the fiscal year. The lag could vary from 9 to 18 months. In the example which follows and which is used to demonstrate the effect of corporation tax on yield, the assumptions are that the corporation tax rate is 30 per cent and the second is that a tax lag of one year operates.

A system of capital allowances also exists and must be included in this example. Capital allowances for industrial plant, including off-highway plant, are currently allowed to be taken at 25 per cent written down. That is, 25 per cent of the investment can be set against tax initially and in the following year 25 per cent of the remaining capital investment can be set against tax. This process repeats in subsequent years; consequently, the amount of the allowance reduces.

Thus, in our organisation, corporation tax is being paid at 30 per cent, a tax time-lag of one year exists, and capital allowances for investment in plant are 25 per cent written down.

If the organisation proposes to invest in a large item of earthmoving equipment, the net cash flows estimated at present-day prices are as given in Table 6.6.

Corporation tax and capital allowances calculations are shown in Table 6.7, where Column *A* shows the original purchase price of £50,000 and the resale value of £15,000. Column *B* shows the net revenues. Column *C* shows the corporation tax that would be due

Table 6.6 Estimated cash flows

Year	*Purchase price and resale value (£)*	*Operating costs (£)*	*Operating revenue (£)*	*Net revenue (£)*
0	50,000			
1		20,000	45,000	25,000
2		22,000	45,000	23,000
3		24,000	40,000	16,000
4		26,000	40,000	14,000
5		28,000	36,000	8,000
6	(15,000)	30,000	35,000	5,000

Note
Total net revenue = £91,000
Capital invested = £50,000
Resale value = £15,000

Table 6.7 Calculation of net tax cash flows

Year	A	B	C	D	E	F	G
	Purchase price and resale (£)	*Net revenue (£)*	*Corporation tax due on previous year's net revenues 30% (£)*	*Capital allowance 25% written down (balancing allowance) (£)*	*Tax saved (or owed) on allowances (£)*	*Tax paid (£)*	*Net cash flow after tax (£)*
0	50,000			12,500.00	3,750.00	(3,750)	−46,250.00
1		25,000		9,375.00	2,812.50	(2,812.50)	−27,812.50
2		23,000	7,500	7,031.25	2,109.37	5,390.63	17,609.37
3		16,000	6,900	5,273.44	1,582.03	5,317.97	10,682.03
4		14,000	4,800	3,955.07	1,186.52	3,613.48	10,386.52
5		8,000	4,200	2,966.30	889.89	3,310.11	4,689.89
6	15,000	5,000	2,400	2,224.73	667.41	1,732.59	18,267.41
7			1,500	(8,325.79)	(2,497.73)	3,997.73	−3,997.73

on these net revenues if no capital allowances were in operation. The corporation tax in column *C* is time-shifted by one year to reflect the tax time-lag. Thus, the £7,500 of calculated corporation tax shown in year two is calculated on the net revenues shown in year one in column *B*.

Column *D* shows the capital allowance of £50,000, calculated at 25 per cent written down of the purchase price, and a balancing allowance of £8,325.79, calculated on the resale value. The balancing allowance is to compensate for the £15,000 of capital returning to the business from the resale. Having already received a capital allowance of £43,325.79 (the aggregate of column *D*) on the original £50,000, the organisation now finds that the capital invested was only £35,000, made up of £50,000 − £15,000: hence the balancing allowance.

Column E is the tax saved by the capital allowance, calculated as the capital allowance times the tax rate. The tax due on the balancing allowance is also calculated at the balancing allowance times the tax rate.

Column *F* is the tax paid, which is column *C*, less the savings in column *E* or plus the tax due in column E.

Column *G* is the net cash flows after tax. The capital invested in year zero is offset by the tax saving that year. Year 1 is the £25,000 net revenue from column B plus the tax saving from column *F*. The explanation for this is that the tax savings can only be taken if the profits and, hence, the tax due are sufficient to accommodate the savings. If this project were the only project the organisation had, then the profits and the tax in year 1 would be insufficient and the tax saving would have to be rolled forward to be absorbed in later years. However, it was assumed that this was a large organisation, and it is further assumed that this large organisation has sufficient profits to accommodate this saving in years zero or one of this project.

Because this project has given rise to those savings (cash that would otherwise have left the organisation), the savings are credited to this project. Years 2–5 are the net revenues from column *B*, less the tax in column *F*. Year 6 is the resale value plus the net revenue from column *B*, less the tax in column *F*. Year 7 is the tax from the column *F*.

The introduction of corporation tax and capital allowances greatly reduces the profit but it also distorts the cash flows. The cash flow in year 1, for example, is larger than before tax considerations, while the others are much smaller. The calculations in Tables 6.8 and 6.9 compare the yields obtained before and after tax.

The results are:

Payback period	2.13 years
DCF yield	31.43%

Before calculating the yield after tax, it is necessary to adjust the cash flows to remove the negative cash flow in year 7. The method of calculation explained previously is known as the single rate calculation and is not suitable for cash flows that have negatives anywhere other than the starting years. If used on cash flows with large negative cash flows at the end, it could produce more than one interest rate, giving a zero NPV. To overcome this, provision is made for the negative cash flow by taking a suitable amount from the previous year's positive cash flow. The amount set aside is the negative cash amount discounted by one year. The rate used to discount the negative cash flow is called the 'earning rate' and represents the return that can be obtained from a safe investment. Thus, using an earning rate of 10 per cent, the negative cash flow of £3,997.73 in year 7 can be discounted to £3,633.93 (£3,997.73 × 0.9090, is the present worth factor taken from tables). If this £3,633.93 is

Table 6.8 Calculation of yield before tax

Year	*Cash flow* (£)
0	−50,000
1	+25,000
2	+23,000
3	+16,000
4	+14,000
5	+8,000
6	+20,000

Table 6.9 Calculation of yield after tax

Year	*Cash flow* (£)
0	−46,250.00
1	+27,812.50
2	+17,609.37
3	+10,682.03
4	+10,386.52
5	+4,689.89
6	+14,633.11

deducted from the positive cash flow of £18,267.41 in year 6, then year 6s cash flow becomes £14,633.11. Thus, the £3,633.93 removed from year 6 provides for the negative cash flow in year 7 and the cash flows are now suitable for a single rate DCF yield calculation as before.

The results of the calculations after tax are:

Payback period	2.06 years
DCF yield	27.85%

Thus, although the corporation tax removed considerable amounts from this investment project, the yield rate of return was reduced only from 31.43 to 27.85 per cent. The reason for the yield remaining so high is the distortion in the cash flows brought about by the 100 per cent capital allowance in the first year. This gave the net cash flows a relatively large positive cash flow in year 1. This occurred only because the tax saving arising from the capital allowance was taken and credited to this project. If the organisation were unable, owing to lack of profit, to take immediately the benefit of the tax savings given by the capital allowances and had to delay taking the savings until enough profits were available, the yield after tax would fall. Thus, the benefits of capital allowances are at their greatest when the organisation is trading profitably enough to use the capital allowances immediately.

The example shown here was presented in the form of appraising a proposed investment, the cash flows of which were estimated at present prices. Taxation, of course, would be calculated on the actual cash flows as they occurred. These actual cash flows would be subjected to inflation and would reflect the effects of inflation. The effects of inflation on yield calculations are explained in the next section.

Yield calculations and inflation

If cash flow estimates are made at present-day prices, the yield calculated on these cash flows will not reflect the effects of inflation. An example would be the purchase and hiring of an item of equipment, as shown in Table 6.10.

The yield of this project is 8 per cent, calculated using two trial rates of interest (7 and 9 per cent) and then interpolation, as shown in Table 6.11.

By interpolation:

$$\text{Yield} = 7\% + (9\% - 7\%) \times \left(\frac{216.81}{216.81 - (-214.24)}\right)$$

$$= 7\% + 1\%$$

$$= 8.00\%$$

As the estimates were based on present-day prices, this 8.00 per cent yield does not include any effects of inflation.

Including inflation at the uniform rate of 10 per cent per year on both the operating costs and the operating revenues, the cash flows adjusted for inflation are as shown in Table 6.12.

The revised net cash flows are in fact simply the original net cash flows inflated at 10 per cent per year. The yield of these cash flows, which now include inflation, is calculated as shown in Table 6.13.

Table 6.10 Cash flow estimates for purchasing and hiring an item of equipment

Year	*Cash flows*			
	Investment (£)	*Operating costs (£)*	*Operating revenue (£)*	*Net cash flows (£)*
0	10,000			−10,000
1		3,000	6,000	+3,000
2		3,500	7,000	+3,500
3		4,000	7,000	+3,000
4		4,500	7,000	+2,500

Table 6.11 Calculation of yield

Year	*Net cash flow (£)*	*Present worth factors at 9%*	*Present value (£)*	*Present value factors at 7%*	*Present value (£)*
0	−10,000	1.0	−10,000.00	1.0	−10,000.00
1	+3,000	0.91743	+2,752.29	0.93457	+2,803.71
2	+3,500	0.84168	+2,945.88	0.87343	+3,057.00
3	+3,000	0.77218	+2,316.54	0.81629	+2,448.87
4	+2,500	0.70842	+1,771.05	0.76289	+1,907.23
			NPV −214.24		NPV +216.81

Table 6.12 Cash flows adjusted for inflation

Year	Investment (£)	Operating costs (£)	Inflation adjustment (£)	Revised operating cost (£)	Operating revenue (£)	Inflation adjustment (£)	Revised operating revenue (£)	Revised net cash flows (£)
0	10,000.00							−10,000.00
1		3,000.00	300.00	3,300.00	6,000.00	600.00	6,600.00	+3,300.00
2		3,500.00	735.00	4,235.00	7,000.00	1,470.00	8,470.00	+4,235.00
3		4,000.00	1,324.00	5,324.00	7,000.00	2,317.00	9,317.00	+3,993.00
4		4,500.00	2,088.45	6,588.45	7,000.00	3,248.70	10,248.70	+3,660.25

Table 6.13 Calculation of yield on cash flows adjusted for inflation

Year	Revised net cash flows (£)	Present value factors at 19%	Present value (£)	Present value factors at 18%	Present value (£)
0	−10,000.00	1.0	−10,000.00	1.0	−10,000.00
1	+3,300.00	0.84033	+2,773.09	0.84745	+2,796.59
2	+4,235.00	0.70616	+2,990.59	0.71818	+3,041.49
3	+3,993.00	0.59341	+2,369.49	0.60863	+2,430.26
4	+3,660.25	0.49866	+1,825.22	0.51578	+1,887.88
			NPV −41.61		NPV +156.22

By interpolation:

$$\text{Yield} = 18\% + (19\% - 18\%) \times \left(\frac{156.22}{156.22 - (-41.61)}\right)$$

$$= 18\% + 1\% \times 0.79$$

$$= 18.8\%$$

This yield, calculated on the revised cash flow, is larger than the yield calculated on the original cash flows, because the revised positive cash flows themselves were larger. These cash flows were larger because of the inflation included in them. The £3,993 revised net cash flow in year 3 does not buy any more goods than the £3,000 original net cash flow. The difference between the two amounts is cancelled by inflation. The £993 more in the revised cash flows just compensates for the inflation at 10 per cent per year. Thus, although the revised positive net cash flows are larger and, hence, the yield is larger at 18.8 per cent, the investor is no better off, because the extra monies acquired are absorbed by inflation. Thus, 8 per cent on the original *uninflated* cash flows is equivalent to 18.8 per cent on the revised *inflated* cash flows. To distinguish between these two rates of return, the rate of return calculated on the estimates which did not include inflation, the original *uninflated* cash flows, is called the *real rate of return*. The rate of return calculated on the estimates which include inflation, the revised *inflated* cash flows, is called the *apparent rate of return*.

The relationship between these two rates of return is through the inflation rate. The apparent rate of return is the real rate of return, increased by the inflation rate as follows:

$$(1 + a) = (1 + r)(1 + d)$$

where a is the apparent rate of return, r is the real rate of return and d is the inflation rate.

Substituting the values calculated from Tables 6.11 and 6.13 for a and r ($a = 0.188$ (18.8 per cent) and $r = 0.08$ (8 per cent)) into the expression gives a value for d, the inflation rate:

$$(1 + 0.188) = (1 + 0.08)(1 + d)$$

$$\therefore \quad (1 + d) = 1.188/1.08$$

$$\therefore \quad d = 1.10 - 1 = 0.10 = 10\%$$

Given the value of the calculated real rate of return at 8 per cent and the calculated apparent rate of return at 18.8 per cent, the expression estimates that the inflation rate is 10 per cent. Since the inflation rate included in the estimates was 10 per cent, this serves as a check on the above explanation.

As most proposed investments are appraised on estimates based on present-day prices, the rate of return calculated and used to judge a proposal's viability is normally the real rate of return. Since the cash flows recorded as projects proceed are normally based on the transactions that occur at current prices, the cash flows determined usually have inflation included as a matter of course. Thus, the yield calculated on these recorded cash flows will be the apparent rate of return. It is, therefore, important to distinguish whether the cash flows are based on constant, year zero, prices or current prices when interpreting the yield calculated.

The situation used to explain the relationship between real rate, apparent rate and the inflation rate was one where the equipment owner was allowed to increase revenue at the same rate as inflation. However, if this were not the case, the apparent rate would be less than 18.8 per cent but the inflation would still be 10 per cent and the *achieved* real rate of return would be less than the *estimated* real rate of return of 8 per cent.

The following example illustrates this. If the project originally described in Table 6.10 had been executed and completed, the cash flows could have been recorded. Throughout the duration of the project, inflation had been 10 per cent per year. Public spending cuts, high interest rates and a general recession had made it impossible for equipment hirers to raise the hire rates in line with inflation, although rates had increased to some extent. Thus, the operating revenue, although increasing, had not kept pace with inflation. The recorded cash flows are as shown in Table 6.14. The apparent rate of return calculated on these recorded cash flows is 12.85 per cent, as calculated in Table 6.15.

Table 6.14 Cash flows recorded during the execution of the project

Year	*Investment (£)*	*Operating costs (£)*	*Operating revenue (£)*	*Net cash flow (£)*
0	10,000.00			−10,000.00
1		3,300.00	6,435.00	+3,135.00
2		4,235.00	8,057.09	+3,822.09
3		5,324.00	8,747.50	+3,423.50
4		6,588.45	9,569.75	+2,981.30

Table 6.15 Calculated apparent rate of return on recorded cash flows

Year	*Recorded net cash flows (£)*	*Present value factors at 13%*	*Present value (£)*	*Present value factors at 12%*	*Present value (£)*
0	−10,000.00	1.0	−10,000.00	1.0	−10,000.00
1	+3,135.00	0.88495	+2,774.32	0.89285	+2,779.08
2	+3,822.09	0.78314	+2,993.23	0.79719	+3,046.93
3	+3,423.50	0.69305	+2,372.66	0.71178	+2,436.78
4	+2,981.30	0.61331	+1,828.46	0.63551	+1,894.65
			−31.33		+177.44

By interpolation:

$$\text{Apparent rate of return} = 12\% + (13\% - 12\%) \times \left(\frac{177.44}{177.44 - (-31.33)}\right)$$

$$= 12\% + 1\% \times 0.85$$

$$= 12.85\%$$

Thus, the achieved apparent rate of return is 12.85 per cent, but as inflation during this time was 10 per cent per year, the real rate of return achieved is 2.59 per cent, calculated as follows:

$$(1 + a) = (1 + r)(1 + d)$$

$$(1 + 0.1285) = (1 + r)(1 + 0.1)$$

$$\therefore \quad (1 + r) = 1.1285/1.1$$

$$\therefore r = 1.0259 - 1 = 0.0259 = 2.59\%$$

Thus, because the income or revenue was restrained from increasing at the rate of inflation but costs were rising at the rate of inflation, the real rate of return was reduced to 2.59 per cent.

Other examples could be apparent rates of return of 10 per cent, inflation rates of 10 per cent and real rates of return of zero, or an apparent rate of 8 per cent, an inflation rate of 10 per cent and a real rate of return of −1.8 per cent.

Inflation evidently reduces the real rate of return unless prices are allowed to rise to compensate for its effects. This reduction in the real rate of return could drag the achieved real rate of return below the cost of capital. In other words, the capital could well be costing more than the project is yielding. In such a case the project is uneconomic, and if such projects are sustained, the business will become bankrupt. The first noticeable effect will probably be that the organisation cannot replace its equipment and will either go on using ageing equipment or reduce its fleet.

The 'construction' equipment hire industry is particularly vulnerable to the effects of inflation. This is because the construction industry is largely an industry that does not create its own demand and is dependent on public works and public spending for a substantial portion of its workload. The rest of the industry's workload depends on the private sector being willing to invest. If the cost of money, that is, interest rates, is high, the private sector is discouraged from investing and the demand for construction work declines. Since in times of rising inflation governments cut public spending while also increasing interest rates, the demand for construction declines accordingly. As the demand declines, prices for construction work fall and the equipment hire industry cannot raise prices in line with inflation. As a result the gap between operating costs and revenue is squeezed, with a consequential fall in the return on capital invested in the equipment.

Methods of acquisition

Much of Chapter 5 and the section on economic profitability (earlier in this chapter) dealt with the economic analyses that help decide whether or not to acquire an item of equipment, based mainly on the question: Does this proposed acquisition offer the opportunity to earn an adequate rate of return and which of the possible items of equipment is the most economic? So far the question of how the item of equipment should be acquired has not been considered. There is a tendency in very large organisations for the two decisions as to whether to acquire and how to acquire to be taken separately, the specialist to acquire responsible for the decision as to whether to acquire and the finance directors being responsible for the decision as to how to acquire. In smaller organisations these two decisions often get merged. Major methods of acquisition are reviewed in this chapter, and the relative advantages of each method are highlighted.

The decision to acquire an asset should be made for both technical and economic reasons. The profitability of the proposal should be evaluated by calculating the expected rate of return and comparing it with the cost of capital. The decision as to how to acquire the asset can then be regarded as a financial one.

The major methods of acquisition can be classified as purchase, leasing or hiring. The major factors that influence the decision as to which is the more advantageous are:

1 Tax legislation, which allows 25 per cent written down capital allowances against the purchase of off-highway plant.
2 The profit flows of the acquiring organisation, which determine whether these allowances can be turned into tax savings benefiting the organisation immediately or rolled forward until later years, thus becoming devalued.
3 The acquiring organisation's cash flows, which determine what money is available for plant acquisition.
4 The acquiring organisation's gearing ratio (borrowed capital/equity capital), which influences the amount of further borrowing possible.

Purchase

Outright purchase

Outright purchase is simply payment of the purchase price by the acquiring organisation to the supplier. This involves the acquiring organisation in a large cash payment very early, before the equipment acquired has earned any revenue. However, outright purchase provides the acquiring organisation with capital allowances of 25 per cent written down of the purchase price of the equipment. If the acquiring organisation's profit flows are sufficient, these allowances can produce a saving of 30 per cent of the allowances. Thus, an item of plant worth £10,000 would produce capital allowances of £2,500 immediately and tax savings of £750.00 (£2,500 × 30 per cent, 30 per cent being the corporation tax rate). This tax saving is most valuable but it is only available if the profit flows in the business are £2,500 or more.

If the cash is available from within the organisation's own resources or even from an overdraft, this form of acquisition is probably the cheapest, provided that the capital allowances can be used to produce the tax saving immediately. If the capital allowances cannot be used immediately because the organisation's profit flows are inadequate, then the capital allowances can be rolled forward until sufficient profit flows are available. In this situation the benefit of the capital allowances and derived tax savings become devalued, in simple present worth (or value) terms, and in these circumstances outright purchase may not be the cheapest method of acquisition.

Outright purchase places the title of the equipment immediately with the acquiring organisation. This means that it becomes an asset over which the organisation has full control; which it can use in negotiating finance; which it can use anywhere, including overseas; and which it can dispose of to produce cash from its resale value.

Credit sale

A credit sale is a sale in which the acquiring takes the ownership or title of the plant item immediately but the purchase price is paid in instalments. These instalments include the purchase price plus any financing charges the vendor makes. Credit sales, like outright purchase, attract capital allowances immediately and can be used in the same way as outright purchase.

Hire purchase

Hire purchase and leasing are significantly different in the treatment of tax and therefore must be considered quite separately. Hire purchase is a contract whereby the acquiring organisation pays a regular hire charge and, at some predetermined point after payment of a proportion of the agreed hire charges, the acquiring organisation buys the equipment for a nominal sum. This facility to purchase distinguishes the hire purchase contract from leasing, which, under UK tax legislation, does not permit the acquiring organisation to purchase the leased equipment. Hire purchase also attracts the capital allowances as though the equipment were purchased outright. Thus, in terms of tax savings, hire purchase has the same advantages as outright purchase.

Both hire purchase and credit sales are likely to require deposits, but these deposits are much less than the whole purchase price and therefore in cash flow considerations these forms of acquisition are less demanding than outright purchase. However, the interest charges included in hire purchase contracts are likely to be greater than those the acquiring organisation would pay on an overdraft. Thus, if the hire purchase method of acquisition is compared with outright purchase, outright purchase is cheaper in most cases, the capital allowances available in both cases being the same.

Leasing

The leasing method of acquisition is different in concept from the previous methods outlined. The difference is that the ownership or title of the equipment remains the property of the leasing company (the lessor), and the acquiring organisation (the lessee) never becomes the owner. The acquiring organisation (the lessee) only acquires the *use* of the equipment in return for payments or rentals but never becomes the owner. While this method is more common in the leasing of property, it is also used in the acquisition of the use of capital equipment such as off-highway plant and equipment. Although there is a plethora of leasing arrangements, they all adhere to the basic principle that the lessor is the owner and the lessee is the user of the equipment. There are two broad categories of lease – the finance lease and the operating lease.

Finance lease

The finance lease is normally arranged through leasing companies who have no particular interest in the equipment and offer no technical support, but merely arrange the lease. The lessee pays the lessor payments or rentals for the use of the equipment acquired. The equipment is usually supplied by a third party – the equipment manufacturer or manufacturers' agent from whom the equipment will be bought by the leasing company. The payments or rentals for this type of lease will be divided into two parts – the primary period and the secondary period. The duration of the primary period is dependent on the useful life of the equipment, but is 2–5 years for most off-highway equipment. Payments during this primary period are calculated by the leasing company to include the capital cost of the equipment, less any allowance for the resale value, plus the leasing company's additions for their own overheads, interest charges and profit. The capital cost will also reflect the capital allowances that the leasing company will be able to claim against the leasing company's tax. Thus the capital allowances due do not go to the acquiring organisation using the equipment, as in purchasing, because the lessee is not the owner. The capital allowances are claimed by the owner who purchased the equipment, and that is the leasing company. The advantages the lessee gains are from the amount of this benefit that is passed on by the leasing company through the payments or rentals. Payments for any secondary period are negotiated but could be relatively small, as the leasing company has already recovered all costs and profit during the primary period. The using organisation is not permitted by tax legislation in the UK to buy the equipment or to sell it, although the leasing company, which is unlikely to have any interest or ability to use the equipment, is free to sell it and may share the proceeds with the user organisation who had leased the equipment. The conditions of such a lease agreement tend to prevent the company that leases the equipment from cancelling the agreement during the primary period. The contract is also likely to specify responsibility for insurance, maintenance, servicing and repairs. All these conditions are designed to protect the owner's property for the duration of the primary period. The leasing company's contribution is simply that of providing finance.

The outgoings of the lease to the user organisation, the lessee, are the lease payments or rentals: the lessee does not have to find the purchase price, or deposits, as in hire purchase or credit sales, but may have to pay about months' rental in advance. Thus, the cash flow for leasing is less difficult to arrange. Some leasing companies have been known to arrange uneven lease payments which are negotiated to match the cyclical use of equipment. Thus, the leasing of earthmoving equipment which may be idle during December, January and February could well benefit from such uneven repayment schemes.

As described the lessee does not receive the capital allowances directly: these go to the owner – the leasing company – and the benefit is passed on via the lease payments.

This benefit is available to the lessee regardless of his profit flows. Thus, whether the lessee (the organisation) has large profits able to benefit from capital allowances or small profits unable to benefit from capital allowances is no longer relevant, as it is in purchasing. Thus, leasing in terms of costs when compared with the purchasing option is more advantageous to an organisation when its profit flows are small and it is unable to use the capital allowances to generate tax savings immediately; or when the organisation is rapidly expanding and its investment programme has created such a total of capital allowances that, even though it is profitable, it is unable to use all these capital allowances. In addition to the capital allowances, the lease payments themselves are normal trading expenses and therefore deductible from revenue before calculating tax due. Thus, there is some tax saving on the lease payments as well as any benefit from the capital allowances the leasing company may have passed on when calculating the lease payment.

Another feature of finance leases is that the security of the lease may well be only the asset itself and the rest of the organisation's borrowings against the organisation's owned assets may not be affected by a leasing arrangement. Therefore, an organisation that is already 'highly geared', that is, has high borrowings in relation to equity or shareholders' capital, may find leasing attractive. Although the leased asset may not show on the balance sheet, the business has nevertheless committed itself to payments and in practical terms is just as vulnerable as if it had increased its borrowings.

Thus, a finance lease is likely to be more advantageous to an organisation when its profit flows do not generate the full benefit of capital allowances for purchasing and/or when the business cash flow situation is unable to provide funds for purchase, or when the organisation is unable to undertake further borrowings to purchase equipment.

Operating lease

The operating lease is normally arranged with manufacturers or suppliers who offer such a service as part of the marketing of their products. Again such leases are likely to have a non-cancellable primary period, but the duration and costs may be quite different from those of finance leases because the leasing company, being the manufacturer, has a different interest in the equipment. For example, the supplying organisation may have use for the equipment itself or a well-developed second hand leasing market. In these circumstances an operating lease may be cheaper than a finance lease. The capital allowances would, as with finance leases, go to the leasing company who own the equipment.

Hiring

Hiring and leasing are sometimes regarded as similar and for equipment items that are on hire for long periods the difference between the two methods of acquisition may be unclear. An example of such a long-term hire is the contract hire arrangements for vehicles. The payments are similar to those in leases, and the owner of the vehicle – the hire company – claims the capital allowances. However, contract hire arrangements can involve the hire company supplying the vehicles in providing repair and maintenance, whereas finance leases do not involve the leasing company in providing these services. Thus, there are distinctions. Short-term hire of equipment is not usually regarded as leasing, and the organisation hiring the equipment pays an hourly, weekly or monthly rate for the equipment. The period of hire may well be as short as one week or one month and therefore the organisation is not committed to a long primary period, as it would be in a finance lease. The use of such short-term hire is, of course, widespread in the off-highway plant and equipment industry, and within the industry there exists a very well-developed equipment hire industry to serve this market. Many organisations who own their own equipment run the equipment division as a subsidiary offering external hire to other organisation and internal hire to their own organisation, which may be the parent company or another company within the same holding group. Thus, the companies, themselves are the users of the equipment, are well used to hiring either internally or externally. This conveniently separates the problems of equipment acquisition, determining adequate hire rates and marketing of equipment to ensure adequate utilisation. All these costs are simply reflected in the hire rate to the organisation.

However, it is worth noting that if a subsidiary equipment hire company sets its internal hire rates at strictly economic levels, internal hirers may be able to obtain ostensibly better external hire rates from outside organisations. These external rates may not have been chosen for economic reasons, or the external hire company may have lower fixed costs and overheads and therefore be able to offer cheaper hire rates. If the external hire is chosen in preference to an apparently more expensive internal hire rate, the parent company may be damaged by the loss of hire income and by the unrecovered part of the fixed costs, which still must be met even while the equipment is idle.

Thus, most of the acquisition of off-highway plant and equipment, whether by purchase or lease, is by the equipment hire companies, which are either companies specifically set up to provide the hire service to organisations or equipment subsidiaries of other organisations.

Comparison between leasing and purchasing

The key to deciding whether leasing is more advantageous than purchasing in cost terms is the use organisations can make of the capital allowances. For example, ignoring tax considerations, if the purchase price of an item of equipment were £12,000 and the lease payments were £1,450 per quarter for three years, the cost of these two cash flows could be compared on present worth terms, using an interest rate that represented the value of money to the organisation. Table 6.16 illustrates this, using an interest rate of 15 per cent.

$$\begin{aligned} \text{Present worth of purchasing} &= -£12{,}000 \\ \text{Present worth of leasing} &= -£1{,}450 \times 9.63496 \\ &= -£13{,}970.69 \end{aligned}$$

Note: The factor 9.63496 is the present worth factor for a uniform series calculated as

$$\frac{(1+i)^n - 1}{i(1+i)^n}$$

where n is the number of periods, 12, and i is the interest rate per quarter. The interest rate per annum was given as 15 per cent. Thus, the interest rate per quarter is calculated, using the compound relationship, as:

$$\begin{aligned} &(1 = i_{\text{quarter}})^4 = (1 + i_{\text{year}})^1 \\ \therefore\ &(1 + i_{\text{quarter}})^4 = (1.15)^1 \\ \therefore\ &i_{\text{quarter}} = \sqrt[4]{(1.15)} - 1 = 0.0355 = 3.55\% \end{aligned}$$

Table 6.16 Comparison between leasing and purchasing, ignoring tax considerations

Year	*Quarter*	*Period*	*Purchasing (£)*	*Leasing (£)*
0		0	−12,000	
1	1	1		−1,450
	2	2		−1,450
	3	3		−1,450
	4	4		−1,450
2	1	5		−1,450
	2	6		−1,450
	3	7		−1,450
	4	8		−1,450
3	1	9		−1,450
	2	10		−1,450
	3	11		−1,450
	4	12		−1,450

Table 6.17 Cash flows and NPV calculation for outright purchase including tax saving from capital allowances used immediately

Year	*Purchase price (£)*	*Capital allowances (£)*	*Tax saving (£)*	*Net cash flow (£)*	*Present worth factors (15%)*	*Present worth (£)*
0	−12,000	3,000.00	900.00	−11,100.00	1.0	−11,100.00
1		2,250.00	675.00	+675.00	0.86956	586.95
2		1,687.50	506.25	+506.25	0.75614	382.79
3		1,265.62	379.68	+379.68	0.65751	249.64
					NPV	−9,880.62

Note

Capital allowances and tax savings after year 3 have been ignored. To include these would reduce the net present value calculated and make the purchase option appear even more economic.

Thus, at an interest rate of 15 per cent the purchase alternative is cheaper. In fact, the interest rate in this example would need to be 27.2 per cent per year or 6.2 per cent per quarter before the leasing costs would just equal the purchase price. The situation is influenced, however, when tax considerations are included.

The leasing versus purchasing comparison is a financial appraisal and usually the discount rate used in these comparisons represents the cost of borrowing. The cost of borrowing is the nominal interest rate, less tax for the profitable business.

Table 6.17 shows the cash flows for outright purchase, using the 25 per cent written down capital allowances immediately and assuming a tax lag of one year to represent the delay in meeting a tax liability or, as in this case, recording a tax saving. The tax saving is treated the same as an inward cash flow, since this has the same effect as a saving that prevents an outward cash flow. The tax saving is calculated on a tax rate of 30 per cent. The NPV at 15 per cent is −£9,880.62, as shown in Table 6.17.

Table 6.18 shows the net cash flows for the leasing alternative. No tax savings from capital allowances are shown, as these would already be reflected in the lease payment charged to the acquiring organisation. However, tax savings derived from the lease payments are shown, as these are normal trading expenses and are deducted from revenue before tax. A tax lag of one year is assumed before the tax reductions are included. The tax rate is 30 per cent. The NPV calculated at 15 per cent is −£10,516.11.

Thus, as with the case of Table 6.16, which ignores tax considerations, when tax considerations are included and the capital allowances available in outright purchase are used immediately, the outright purchase option is cheaper than the lease option, as the net present values of −£9,880.62 for outright purchase and −£10,516.11 for leasing illustrates.

However, the situation changes if there is a delay in taking the tax savings from the capital allowances created by the purchase.

Table 6.19 calculates the NPV for a delay to year two before taking the tax savings. This delay could be caused by profit flows being insufficient to use the capital allowances immediately.

With a delay to year two before the tax savings become effective, the difference between purchasing and leasing narrows, and the NPV of leasing remains at −£10,516.11, whereas the NPV of purchasing becomes −£10,176.65.

Any further delay in taking the tax savings will make leasing a cheaper alternative, as Table 6.20 illustrates by adding a further year delay before taking the tax saving.

Thus, with a delay to year 3 before the tax savings become effective, the leasing option becomes more economic as the NPV of purchasing becomes −£10,381.92, whereas the NPV of leasing remains at −£10,516.11.

Table 6.18 Net cash flow for the leasing alternative including tax considerations

Year	*Quarter*	*Period*	*Lease payment (£)*	*Tax reduction (£)*	*Present worth factors (15%)*	*Present worth of tax reduction (£)*
	1	1	−1,450			
	2	2	−1,450			
1	3	3	−1,450			
	4	4	−1,450			
	1	5	−1,450			
	2	6	−1,450			
2	3	7	−1,450			
	4	8	−1,450	+1,740	0.75614	+1,315.68
	1	9	−1,450			
	2	10	−1,450			
3	3	11	−1,450			
	4	12	−1,450	+1,740	0.65751	+1,144.06
4	4	16		+1,740	0.57175	+994.84
The tax reduction is calculated as 30% of a year's lease payments and delayed 1 year (4 × £1,450.00 × 30%).						
Present value of tax reductions						+£3,454.58
Present value of lease payments, from Table 6.16						−£13,970.69
NPV						−£10,516.11

Table 6.19 NPV of outright purchase with a delay to year 2 in using capital allowances

Year	*Purchase (£)*	*Tax saving (£)*	*Net cash flow (£)*	*Present worth factors (15%)*	*Present worth (£)*
0	−12,000		−12,000	1.0	−12,000
1					
2		+2,081.25	+ 2,081.25	0.75614	1,573.71
3		+379.68	+ 379.68	0.65751	249.64
				NPV	−10,176.65

Note
The tax savings are calculated by rolling forward the unused allowances calculated in Table 6.17.

Table 6.20 NPV of outright purchase with a delay to year 3 in using capital allowances

Year	*Purchase (£)*	*Tax saving (£)*	*Net cash flow (£)*	*Present worth factors (15%)*	*Present worth (£)*
0	−12,000		−12,000	1.0	−12,000
1					
2					
3		2,460.93	+2,460.93	0.65751	1,618.08
				NPV	−10,381.92

Note
The tax savings are calculated by rolling forward the unused allowances as calculated in Table 6.17.

Table 6.21 NPV of outright purchase with capital allowances used in years 2 and 3

Year	*Purchase (£)*	*Tax saving (£)*	*Net cash flow (£)*	*Present worth factors (15%)*	*Present worth (£)*
0	−12,000		−12,000	1.0	−12,000
1					
2		1,230.46	+1,230.46	0.75614	930.40
3		1,230.46	+1,230.46	0.65751	809.03
				NPV	−10,260.57

A final illustration shows the capital allowances spread over years 2 and 3 to represent the ability to take some tax savings in year 2 and the remainder in year 3. This is shown in Table 6.21. Again this illustrates that delays in taking the tax savings can lead to the leasing option becoming more economic, as the NPV for outright purchase is −£10,260.57, while the NPV for leasing is −£10,516.11.

In all the lease versus purchase examples shown above, a tax lag of one year was adopted to represent the delay in meeting a tax liability. This one year lag could vary and the lag itself influences this lease versus purchase comparison. However, the point is made in these examples that any delay in using the capital allowances that are created by purchasing shifts the balance of the economic comparison in favour of leasing. Thus, the cost comparison between leasing and buying is determined by the organisation's profit flows.

The other factors to be considered are:

1. The organisation's cash flow, since leasing makes fewer demands than purchasing.
2. The organisation's ability to raise the capital to purchase, since this depends on the extent of the organisation's existing borrowings.
3. The availability of alternative uses for the investment funds available.
4. The loss of flexibility, due to commitments, to make lease payments annually.
5. The restrictions placed on the using organisation in using the equipment.

Thus, the resolution of the lease versus purchase issue is not simply an economic comparison but involves these other factors as well.

Chapter 7

Selection of equipment and hire rate calculation

Introduction

Generally equipment is either purchased for permanent ownership or alternatively hired/leased for the required temporary period to meet peak demand or specialised duties. The specific decision to purchase will have important financial consequences for the firm, since considerable capital sums will be locked up in the item, which must then be operated at an economic utilisation level to produce a profitable rate of return on the investment. When examining the need to own equipment the following points must be considered:

1 Will the item generate sufficient turnover to provide an adequate rate of return on the capital employed?
2 Is ownership, rather than obtaining it by some other method, absolutely necessary for the business?
3 Is outright purchase the only way of acquiring the plant?

If the objective answer to these questions is confident and positive the purchase decision can go ahead and thereafter a realistic rental charge be determined for recovering the ownership and operating costs over the planned life of the asset. If the criteria cannot justify the decision, then there needs to be some other sound commercial reason for making the purchase.

Technical evaluation

The purchase of an item of equipment requires a high capital outlay, and the consequences of misjudging the potential earnings of the machine over a number of years could have a dire effect on future profits. Thus, unless it can be shown that the equipment will yield a rate of return at least as good as making an alternative investment, it should not be purchased at all. In this event either leasing or hiring may be a more economical option. However, even when the financial and other economic factors have been adequately satisfied regarding the purchase, because of the many alternative choices of manufacturer now available for many items of equipment, the final decision is likely to be influenced by the merits of small but important differences offered with each make. It is this aspect of the purchase, namely the technical evaluation that is dealt with first.

The decision process follows a systematic i.e. 'value management' approach originally developed by US consultants Kepner and Tregoe. The method forces the manager into a sequence of actions and helps to highlight the relevant factors. In this way the many separate

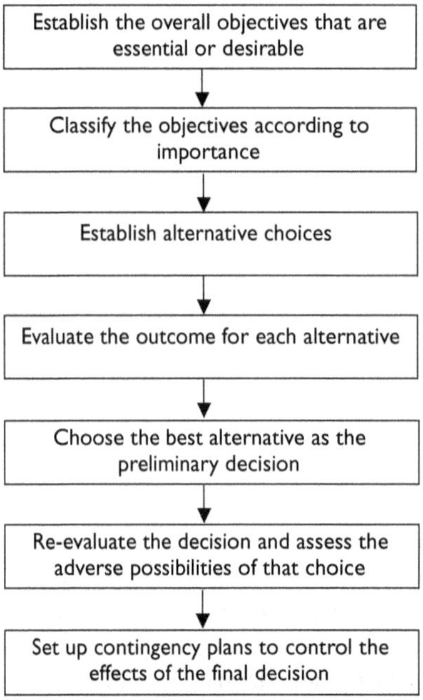

Figure 7.1 Seven essential factors in decision making.

judgements needed for an examination of many facts can be weighted and ranked accordingly, and the best buy for the least cost can be chosen.

The main features of decision making

Dixon describes decision making as follows:

> Decision making is compromise. The decision maker must weigh value judgements that involve economic factors, technical practicabilities, scientific necessities, human and social considerations, etc. To make a 'correct' decision is to choose the one alternative from among those that are available which best balances or optimises the total value, considering all the various factors.

Using this definition, Kepner and Tregoe established seven essential factors in decision making (Figure 7.1).

Example 7.1: Crane evaluation

A company has recently obtained a contract that will involve the use of a 40-tonne-capacity, strut-boom, crawler-mounted crane. A nine-year working life can be assumed, which indicates that an attractive rate of return may be achieved by purchasing the crane. The crane must be capable of conversion to a dragline-grabbing excavator and cost less than £41,000. Management is required to undertake a technical evaluation of the alternative makes of crane available in the market.

Purpose of decision: evaluation of cranes for possible purchase and addition to company fleet

Objectives		*Alternatives*	
Ref	*Musts*	*Harris 46T*	*Go/No*
A	Crawler mounted machine	Yes	✓
B	At least 50 m boom as lift crane	60 m max main	✓
C	Capable of conversion to dragline/excavator	Yes	✓
D	Delivery less than 10 weeks	6/8 weeks	✓
E	Max purchase price £41,000	Yes	✓
F	Confirm to B.S. and statutory regulations	Yes	✓
G	40 tonnes minimum lifting capacity	Yes	✓

No.	*Wants*	*Element*	*Rk*		*Information*	*Rt*	*Weighted score = Rk = Rt*	
1	Low purchase price	S	10		£40,150	9	90	
2	Good lifting capacities	T		10	Max 46.0T	7		70
3	Good dragline capacities	T		7	Max 6T	10		70
4	Good service facilities/ technical banking	S	7			7	49	
5	Low running costs	S	8		No	—	—	—
6	Low upkeep cost/robust construction.	S/T	4	4	information	7	28	28
7	Good safety features. Reliable safe load indicator.	T		8	Weight load	10		80
8	Low machine weight/cost of transport between site	S	2		55T	10	20	
9	Low ground bearing pressure	T		3		10		30
10	Reliable power unit/smooth operation	T		7	9.06 psi (ave) Rolls-Royce	10		70
11	Fast slewing speed	T		5	diesel	10		50
12	Fast hoist speed (single line)	T		6	3.18 rpm	10		60
13	Fast travel around site	T		4	150 f/min	9		36
14	Familiarity with machine	S	4		0.750 mph	6	24	
15	Manoeuvrability/compact geometry/travel on inclines	T		5	Max grad 1 in 4	10		50
16	Long working life/good trade-in price	S	8		9 years – £3,000	10	80	
17	Short rigging time/ease of conversion to dragline	T		4		7		28
18	Operator view and comfort/ good layout of controls	T		5		7		35
S	Cost and service element	39%	43	68			291	607
T	Technical element	61%						
	Totals	100%	111				898	

Figure 7.2 Decision analysis sheet.

Notes
Rk = Ranking.
Rt = Rating.

Alternatives				*Adverse consequences*			
HFC 714	**Go/No**	**Sirac Highlift**	**Go/No**	**Harris 46T**	**P**	**I**	**P3I**
Yes	✓	Yes	✓	Strike may		£	£
50 m max main	✓	60 m max main	✓	delay	0.3	1,200	360
Yes	✓	Only released as lifting crane	×	delivery			
6/8 weeks	✓	8 weeks	✓				
Yes	✓	Yes	✓				
Yes	✓	Yes	✓				
Yes	✓	Yes	✓				

Information	*Rt*	*Weighted score*		
£39, 482	10	100		£36, 534
Max 53.5T	10		100	Max 45T
Max 6T	8		56	
	10	70		
No information	—	—	—	No information
	10	40	40	
Wylie	9		72	Wylie
66T	9	18		50T
9.50 psi (ave)	9		27	8.50 psi (ave)
Dorman diesel	7		49	Dorman diesel
3.00 rpm	9		45	4.00 rpm
140 ft/min	9		54	156 ft/min
0.818 mph	10		40	0.907 mph
10	40			
Max grad I in 4	9		45	max grad I in 4
9 yrs – £3,000	10	80		
10		40		
10		50		
		348		This crane is removed from the analysis as must C is violated
		618		
		966		

Information			
HFC 416	**P**	**I**	**P×I**
As a result of design fault on the clutch some down-time may be necessary to effect a repair		£	£
	0.9	300	270

Figure 7.2 (Continued).

The method

Step 1: Set objectives which are essential and desirable

The problem as defined is far too vague to establish the precise requirements, and some discussion with the project manager, plant manager, planning staff and preferably the operator also is required. From these further enquiries the precise features of the machine may be ascertained: these will be called the 'objectives', and examples are listed A–G, 1–18 in Figure 7.2.

Step 2: Classify objectives according to importance

A careful examination of the listed objectives will reveal that some are mandatory; some are important but not essential, while others would be desirable and useful if available. From this list of objectives, the mandatory objectives are set aside and called *musts*. The *musts* set the limits that cannot be violated, so makes of machine that do not meet the fundamental requirements are quickly recognised and eliminated early in the analysis.

The remaining objectives are labelled *wants*, and some may be more important than others, but may include some or all of the *musts* as appropriate for the evaluation. The next step, therefore, is to rank the importance of the *want* objectives. This is achieved by attaching a ranking expressed on a suitable scale. In practice, a 1–5 or 1–10 system is suitable, with the importance of the objective increasing with numerical value. This part of the procedure usually requires much subjective assessment and may vary according to an individual's preference, but at least a record stands for future reference and reconciliation.

The *want* objectives are classified into two groupings:

1 The performance results expected from the machine (here referred to as the *technical element*);
2 The demands made upon resources of labour, money, materials, time (here referred to as the *economic element*), as associated with costs and servicing.

Step 3: Develop the alternative choices

In this arbitrary example, three (entirely fictitious) manufacturers are located who are able to supply a crane which may be suitable for the specified duties. These are: (1) Harris 46T; (2) HFC 416 and (3) Sirac Highlift. At this stage only the *musts* are considered. Consequently, the field of choice may be narrowed to a small number, saving much effort, time and money.

Step 4: Evaluate the alternative against the objectives

The stage is now reached where it is necessary to consider in detail the specification of the selected machines. Difficulties will now begin to emerge: for instance, there is rarely a unique price for a large item of equipment, as each manufacturer will offer features slightly different from those of its competitors' machines.

In this example the quotations received for the three cranes are shown in Table 7.1. It can be seen that the information assembled for each crane varies in its detail, and any decision on price can be only a compromise. However, within the limits imposed by these variations, the relative prices of the machines can be compared. But it must be remembered that

Table 7.1 Evaluating alternative quotations

Component	*Harris*	*HFC*	*Sirac*
Base machine	£32,809	£26,833	£34,070
Cat head boom	(30 ft) inc.	N/A	N/A
Taper top boom	(27 ft) £756	(50 ft) £3,915	inc.
Boom inserts to max.	£1,469	£2,524	inc.
Fly jib	(30 ft) £573	(30 ft) £561	(40 ft) £1,482
Power lowering equipment	inc.	£2,135	inc.
Safe load indicator and tests	£1,830	£969	£902
Cab heater	inc.	£90	£80
Hook block	£213	£181	inc.
Parts to convert to dragline	£2,500	£968	N/A
$2\frac{1}{2}$ yd^3 dragline bucket	inc.	£1,306	N/A
Total cost	£40,150	£39,482	£36,534

discounts and payment arrangements may influence the final decision. Also, the cost and availability of spare parts and maintenance should not be overlooked at this stage, especially from foreign suppliers affected by exchange rate fluctuations.

Table 7.1 reveals that the Sirac Highlift was not quoted with dragline equipment and that, as it is a new model, it can only be used as a lift crane at present. Therefore, although this machine slipped through the preliminary screening of the *must* objectives, it is now marked *no go* against objective C in Figure 7.2 and is eliminated from further consideration. The Harris and HFC cranes comply with all the *must* objectives and effort can now be concentrated upon the *want* objectives of these two machines.

The *want* objectives are now separated into two groups, namely those that can be evaluated from the manufacturer's specification and those that cannot, as follows:

Can		Cannot	
1	Purchase price	4	Maintenance requirements
2	Safe working loads	5	Running costs
3	Dragline range	15	Manoeuvrability
6	Method of assembly	17	Assembly time
7	Safety features	18	Driver comfort
8	Weight of machine		
9	Bearing pressure under tracks		
10	Bearing		
11	Slewing action		
12	Hoist system		
13	Travelling speed		

For those *wants* which are not readily available from manufacturers' information, more subtle sources must be found. For example, site demonstrations, visits to the manufacturer, records of data on experiences with similar machines, personal contacts with other plant users and discussions with machine operators.

With sufficient information on all the *wants* objectives available, the two remaining crane alternatives (namely Harris 46T and HFC 416) are compared for performance. Each *want* objective is considered in turn and rated on a 1–10 scale. The technical element and economic element ratings are separated in order that the importance of each can be ascertained.

To obtain the relative worth of each *want* objective, the *rating* value is multiplied by the previously given *rank* number. The resulting weighted score represents the performance of the crane against its objective. The weighted scores for each objective are subsequently added, to give totals for each crane alternative, and the totals for each provide an indication of the relative position with respect to the specified objectives. In this example the total weighted scores shown in Figure 7.2 are summarised as follows:

	Percentage of total	Harris	HFC
Cost and service element (S)	39	291	348
Technical element (T)	61	607	618
	100	898	966

Note: It must be emphasised that these values are determined largely by subjective judgement, albeit based on careful analysis of the available facts, and as such do not make the selection of the best alternative a routine procedure. However, the values do make it possible to deal systematically with many factors which otherwise could not easily be related for importance.

Step 5: Choose the best alternative as the preliminary decision

The alternative that receives the highest weighted score is presumably the best option; in this case HFC 416. It is, of course, not a perfect choice, for instance, one would probably prefer a machine with a Rolls-Royce engine. But the choice is at least one that strikes a reasonable balance between the good and bad features of the machine. Fortunately, both the technical and cost element totals are greater for the HFC 416. The decision would be far more difficult if one of the elements were larger for the Harris 46T crane. Some further balance would then be needed between economic resources and the technical benefits accruing to the company.

The preliminary choice is the one which best satisfies the objectives overall. It is thus a compromise, as undoubtedly the alternatives have some superior features. But the method used shows the manager how the decision was arrived at and therefore where the possible pitfalls may lie.

Step 6: Re-evaluate the decision and assess the adverse possibilities of that choice

When many alternatives are available, the weighted scores of one or two are sometimes quite close, so, before the final decision is made, any adverse consequences must be considered. The manager must look for snags, potential shortcomings or anything else that could go wrong. The probability (P) of the adverse consequences should be assessed and a seriousness weighting (I) given to its possible effect. An expression of the total degree of threat may be obtained by multiplying the seriousness weighting by the probability estimate.

In this example the plant and site managers arrange a meeting to discuss the implications of the proposals, and the following points are raised.

(1) The discarded Sirac Highlift appeared to be a fairly good crane on preliminary inspection and perhaps the *must* objective necessitating dragline conversion could have been reduced to a major *want* objective. A further check on the records reveals that only 20 per cent of the time was spent on such duties. Thus, unless the plant manager insists that this facility is essential, this crane should be reconsidered. But the *must* objective should only be waived in very exceptional circumstances.

(2) Possible adverse consequences of choosing the Harris 46T, such as the effects of a strike at the factory, can be assessed from past experience. It may be that on the past 10 occasions when a new wage agreement has been under negotiation between the Harris company and its workforce, a strike resulted on three occasions. So the probability (P) of a strike at this time is judged to be 30 per cent. If a strike did occur which would last more than a few weeks, then the delivery of this machine would exceed the specified time on the order, in which case the crane should not be considered at all. However, should a strike occur lasting four weeks, the cost of hiring a crane would be about £300 per week; therefore, the degree of threat is $(4 \times 300) \times 0.3 = £360$.

(3) One possible adverse consequence of the HFC 416 has come to light. A serious accident has just revealed a major design fault in the clutch mechanism of the winch. A temporary

modification will be made to all existing models and stocks held. But a completely new part is not available for six months and will require taking the crane out of service for one week at that time. The probability that this service is required is 90 per cent and the cost of hire for another crane is £300 per week, thereby directly increasing the contract cost at this rate. The degree of threat is $(1 \times 300) \times 0.9 = £270$. The choice, therefore, is still the HFC 416. Had the consequences been reversed, a much closer examination of the advantages and disadvantages would be necessary, making the decision much more difficult.

Step 7: Set up contingency plans to control the effects of the final decision

The adverse consequences represent potential problems, and they must be prevented from causing too much inconvenience. This is done either by taking preventative action to remove the cause or by deciding upon a contingency action if the potential problem occurs. In our case, the simple remedy is to ensure that alternative machines are available for hire at suitable times or to arrange work on site so that interruption is minimal.

Comments

A decision has been reached to buy the HFC 416, and this is the best choice on the basis of the judgement of the plant manager concerned. For someone else with different experience, the assessment of the objectives might be quite different and would possibly produce a different result. But even accepting this obvious weakness, the method at least forces the manager to consider most of the facts. It provides a record of the thought processes and will help to sort out points of confusion, for it is almost impossible to memorise all the facts and relationships, except for the very simple choices. In such cases the decision process can be adequately carried out mentally without the need for the elaborate method described. Finally, the decision process is written down on paper and is available for all to see.

The whole procedure can now be readily performed on a personal computer, using commercial spreadsheet and database packages. Indeed, equipment manufacturers themselves may eventually offer comprehensive information such as machine specifications, performance characteristics, prices, etc., on computer discs, which would considerably assist the analysis process.

Calculating the hire rate/rental charge

Once the equipment has been selected and purchased then revenues from hiring out/renting to the open market or obtained solely from internal usage, should not only recover the owning and operating costs, but also aim to achieve expected profitability.

Ownership costs are fixed costs arising indirectly, such as business overheads. They are incurred throughout the period of ownership and are a fixed charge. They include:

- cost of capital;
- depreciation on the equipment;
- insurances and licences;
- corporation tax and capital allowances;
- establishment charges.

The operating costs are direct costs of material, labour and expenses, which vary according to the usage of the machine, and include:

- servicing costs – for example, oil, grease and other consumables;
- maintenance costs;
- transport charges;
- fuel;
- operators' wages.

Cost of capital

Capital to purchase machinery (a fixed, tangible asset) is generally obtained from borrowing or retained profits. In either case the interest charge on the loan or an acceptable rate of return on private funds should be incorporated into the hire rate.

Depreciation

Most items of equipment wear out and deteriorate with usage or become obsolete with time. Thus, at the end of its economic life sufficient revenues should have been generated to cancel out the capital part of a loan or to replace the machine when internal funds are used. In calculating depreciation, it is necessary to estimate the resale or scrap value of the machine and the useful life. Estimation of the useful life is mostly done by intuition based on experience of operating a similar item. For example, when the utilisation level shows a clear downward trend accompanied by increased maintenance costs and loss of profit, the machine has probably reached the end of its useful life for the organisation.

There are several methods of calculating the annual depreciation charge, but the choice will depend largely upon the type of machine and its operation. In principle, depreciation is deducted from profits (since it represents a cost on the business activities) before tax is paid. It is, therefore, necessary to obtain approval of the method and depreciation period from the Inspector of Taxes before a suitable policy can be adopted. However, where capital allowances are in operation special arrangements generally apply (see Chapters 6 and 16).

Straight line depreciation

The most common method of depreciation is called straight line depreciation. It is also the most simplistic and convenient means of writing off capital expenditure over accounting periods. The technique allows an equal amount of depreciation to be charged for each year of the expected use of the plant item and can be calculated using the following formula:

$$\frac{Dp}{a} = \frac{C - S}{N}$$

where Dp/a is the depreciation per annum; C is the purchase price of the asset; S is the salvage (residual) value of the asset at the end of its useful life; and N is the life expectancy of the asset (in years). This method is, however, misleading because most plant items tend to lose value more rapidly in their early life, followed by a slower rate of depreciation thereafter (Edwards *et al.*, 1998).

Example 7.2

It is decided to purchase a mechanical excavator costing £42,000 (C) to work on average 2,000 h per year. The life of the machine is expected to be 10 years (N), after which the salvage value will be £2,000 (S).

$$\therefore \quad \frac{Dp}{a} = \frac{£42{,}000 - £2{,}000}{10}$$

The total depreciation is £40,000, the annual depreciation is £4,000. The hourly depreciation will therefore be £4,000 ÷ 2,000 (hours operation p/a) = £2.

Declining balance depreciation

Using the data from Example 7.2, the declining (or reducing) balance method calculates the provision for depreciation annually, on the balance of the asset from the previous year. For this method, the depreciation is made on a fixed percentage basis rather than a fixed sum. Declining balance is particularly useful for plant and equipment assets because repair and maintenance costs typically increase as the machine ages. By reducing the provision for depreciation as the repair and maintenance costs rise, the total usage costs each year are kept fairly constant.

The figures in Table 7.2 can more easily be calculated from the formula:

$$d = (1 - N\surd S/C) \times 100$$

where S is the salvage value, C is the purchase price, N is the life of asset and d is the percentage depreciation.

Figure 7.3 demonstrates the main features of the declining balance method and the straight line method.

The declining balance method of depreciation is more representative of a real life scenario since it allocates larger charges of depreciation early in the life of plant. These charges

Table 7.2 Declining balance depreciation example (at 26.2%)

End of year	*Depreciation* (£)	*Depreciation for year* (£)	*Book value* (£)
0	26.2	0	42,000
1	26.2	11,004	30,996
2	26.2	8,121	22,875
3	26.2	5,994	16,881
4	26.2	4,423	12,458
5	26.2	3,264	9,194
6	26.2	2,409	6,785
7	26.2	1,778	5,007
8	26.2	1,312	3,695
9	26.2	968	2,727
10	26.2	727	2,000
		Total: £40,000	

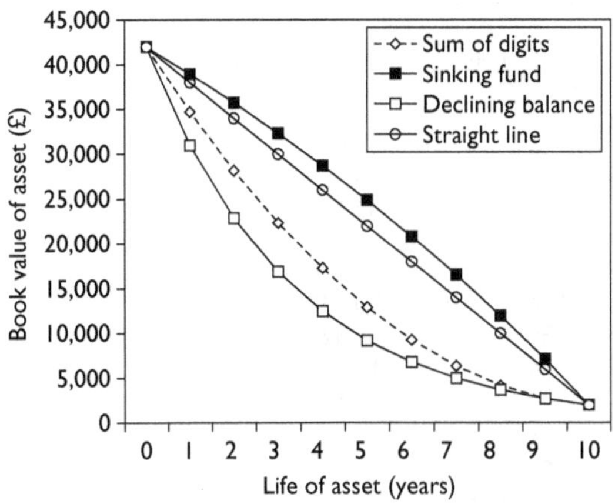

Figure 7.3 Graphical comparison of depreciation methods.

reduce in amount over the remaining life of the equipment. In contrast the straight line method produces a constant (linear) periodic depreciation amount over the asset's life. Because it is simple to calculate, the straight line method tends to be favoured by plant managers with mixed plant fleets, having a range of purchase and disposal dates.

Sinking fund method of depreciation

In this method, a fixed sum is put aside from revenue each year and invested with compound interest throughout the life of the asset. After successive instalments the sum accumulates, to produce the original purchase price less the scrap value.

Taking the given example and assuming that 6 per cent interest is earned on savings, the annual amount required is £3,034, thus:

Uniform series factor that amounts to one over 10 years = 0.07586
Therefore, annual payment = 0.07586 × (£42,000 − £2,000) = £3,034

Table 7.3 gives a detailed breakdown of the analysis.

Sum of digits method

When using the sum of digits (SD) method, an arithmetic sequence is set up based on the total useful life of the machinery. The depreciation percentage rate is calculated to be proportional to the remaining useful life. Information provided in Table 7.4 is based upon a machine life expectancy of 10 years. To obtain a proportional depreciation rate the sum of digits is calculated as follows:

$$\mathrm{SD} = (C - S) \times \frac{\left(\sum \text{digits (years)}\right)}{N}$$

Table 7.3 Sinking fund example (6% interest on savings)

Year	*Payment* (£)	*Interest* (£)	*Depreciation* (£)	*Book value* (£)
1	3,034	0	3,034	38,966
2	3,034	182	3,216	35,750
3	3,034	375	3,409	32,341
4	3,034	581	3,615	28,726
5	3,034	798	3,832	24,894
6	3,034	1,028	4,062	20,832
7	3,034	1,271	4,305	16,527
8	3,034	1,529	4,563	11,964
9	3,034	1,803	4,837	7,127
10	3,034	2,093	5,127	2,000
			Total: £40,000	

Table 7.4 Sum of digits depreciation example

Year	*Factor*	*Depreciation* (£)	*Book value* (£)
1	10/55	7,273	34,727
2	9/55	6,454	28,182
3	8/55	5,818	22,364
4	7/55	5,091	17,273
5	6/55	4,364	12,909
6	5/55	3,636	9,273
7	4/55	2,909	6,364
8	3/55	2,182	4,182
9	2/55	1,454	2,728
10	1/55	728	2,000
		Total: £40, 000	

Therefore, for year 1 (£42,000 (C) − £2,000 (S) = £40,000) × (10 (N) ÷ 55); where 55 is the summation of the years one to 10 (1 + 2 + 3 + 4 + 5 + 6 + 7 + 8 + 9 + 10). The depreciation rate for year 1 is £7,273.

Therefore, for year 2 (£42,000 (C) − £2,000 (S) = £40,000) × (9 (N) ÷ 55). The depreciation rate for year 2 is £6,545.

Therefore for year 3 (£42,000 (C) − £2,000 (S) = £40,000) × (8 (N) ÷ 55). The depreciation rate for year 3 is £5,818.

Free depreciation

The asset is totally depreciated initially by this method. For example, the item of equipment purchased for £42,000 would be depreciated by £40,000 immediately on acquisition, leaving only the £2,000 salvage value.

Licences and insurances

The costs for vehicle excise, driver and goods operator licences largely depend upon the type of vehicle/equipment and whether the equipment is to be used on the public roads or otherwise. Acquisition of appropriate insurance is also necessary to cover risks to third parties, adjacent property and public utilities. These aspects are fully discussed in Chapter 8.

Establishment charges

The owner needs to recover the cost of overheads in the hire rate, that is, the central organisation, including the offices, workshops and associated administrative facilities. These fixed costs are commonly apportioned on the basis of budgeted annual operating hours of the item or as a percentage of the purchase price of the machine per annum.

Operating costs

Operating costs are (as their title implies) directly attributed to the cost of operating the machine and include maintenance, consumables and operator wages. Without such expenditure the many benefits of mechanisation could not be realised.

Maintenance

Equipment requires periodic maintenance to keep it in working order, and facilities are required to provide both a fixed-time-to and a breakdown maintenance service. The budgeted costs of maintenance to include in hire rate calculations should be derived from records of the operating costs incurred by the plant workshop. Indeed, the workshop itself should operate a cost control system for each item of plant, to record the expenditure on spares, maintenance, etc. The costs of maintenance are variable and ought therefore to be included in the hire rate as a direct cost per hour operated. In practice, however, many owners prefer to express the costs of maintenance in the hire rate as a percentage of the purchase price of the machine.

Consumables

Consumables include oil and grease, tyres and fuel, and are direct costs which vary according to the condition of the machine, work done and hours operated. The cost of greases and lubricants incur in two instances. First, they are used during operator maintenance undertaken on a daily basis. Second, larger quantities of oils are consumed during periodic changes at those intervals either recommended by the plant manufacturer or as indicated by a reputable supplier of lubricants. It will be readily appreciated that there will be a great deal of variation in oil usage, according to: machine type; operating hours per week; machine age; and operating conditions.

Consumable costs are infamously difficult to estimate and will vary depending upon the region of purchase, allowable discount and quantity purchased. Accurate costs can however be accrued from records of operating similar equipment in the past. In the absence of such data, manufacturers provide guidelines on the consumption of these materials, but care should be exercised, as the data are likely to relate to new equipment operating under ideal conditions. The inclusion or otherwise of the cost of fuel in the hire rate will, of course, depend upon the hire contract conditions.

Operators' wages

The operators' wages relates to the direct cost of employing a competent person(s) to operate plant items. The actual gross wage paid to the operator is dynamic as it contains so many variables, a combination of which could lead to infinite possibilities. Broadly speaking, a composite rate would be built up of the following constituent parts: overtime, bonus, travelling subsistence, national insurance premiums, holiday pay, sick pay, pensions, etc. Operator wages are not usually included in the hire rate since many items are hired exclusive of the operator. But, when required, costs based on the driver's hourly rate must be recovered; when this occurs careful consideration should be given to developing an appropriate composite rate. The rate should be high enough to attract the best candidate but not too high as this would negatively impact upon competitiveness.

The effect of corporation tax and capital allowances

These are treated more fully in Chapters 6 and 16 but, briefly, depending upon national policy, governments sometimes introduce investment incentives to encourage businesses to account for free depreciation on equipment, so that an item is allowed the full purchase price to be deducted from profits in the year of purchase, and thereby the amount of corporation tax due is reduced in that year. The tax, however, will, of course, be recovered in full from profits in subsequent years, since there is no allowance to set aside thereafter. The effect of this arrangement is to alter the sequence of cash flows for the business, which may produce a more favourable return on capital employed and so encourage other investment.

The effect of capital allowances, when introduced by government, is aimed at encouraging firms to purchase new equipment and boost economic performance. If used to excess such a policy has the potential to cause the market to become oversupplied with plant resulting in hire rates falling below levels which can earn a healthy rate of return on capital employed. Thus, although over the short-term customers and both home and overseas manufacturers benefit, ultimately low profitability cannot be sustained indefinitely and considerable adjustment becomes unavoidable either through bankruptcy or mergers.

It is emphasised that capital allowances are basically a feature of the profit and loss account for the whole enterprise for the calculation of corporation tax payments. Clearly, to allow full initial depreciation in the hire rate would be non-sensical, as the rate would be uncompetitive during that year thus, for such purposes internal depreciation should be based on a realistic assessment of the life of the asset.

Calculating the economic hire rate

There are several acceptable methods of calculating an economic hire rate with the favoured approach requiring calculation of ownership and operating costs plus a contribution for profit. However, a more satisfactory method for investment extending over a few years is the discounted cash flow (DCF) yield, which takes into account the timing of cash flows.

Example 7.3: The conventional method

An excavator, crawler-mounted, of $1\frac{1}{2}\,m^3$ capacity, is purchased new for £46,000. Its estimated life is 10 years, with a resale value of £4,000 (refer to Table 7.5). Hire charge = £18,054/2,000 = £9.03 per hour (or £361 per 40 hour week).

Table 7.5 The conventional method for calculating an economic hire rate

Cost considerations	*Values*
Capital cost	£46,000
Resale value	£4,000
Anticipated life	10 years
Insurance premium	£200 per year
Road tax and licences	£100 per year
Maintenance	10% of capital cost
Consumables	£400 per year
Overheads of business	£2 per hour
Required rate of return on investment	15% per year
Budgeted operating time	2,000 hour per year
Transport charges	say £100
Item	*£ per year*
Depreciation (straight line) $= \frac{42{,}000}{10}$	4,200
Interest on finance, calculated using a capital recovery factor from interest tables (CRF = 0.199 at 15% p.a. for 10 years):	
$\frac{(46{,}000 \times 0.199 \times 10) - 46{,}000}{10}$	4,554
Fixed overheads = 2 × 2,000	4,000
Road tax and licences	100
Insurance premium	200
Ownership (fixed) cost	13,054
Consumables	400
Maintenance × 10% × 46,000	4,600
Operating cost (variable)	5,000
Total cost	18,054

The cost of transport and any additional profit should be added to this figure. Finally, the hire rate is based upon a utilisation period of 2,000 h, and should this target not be reached, the ownership (fixed) costs will be under recovered and budgeted profit will not be achieved.

Alternative analysis using DCF (see Chapter 6)

DCF takes into account the timing of cash flows, whereby income and outgoings are balanced to yield a satisfactory return, in this example 15 per cent per annum. The problem is thus restructured as shown in Table 7.6.

Information used in the table:

1 4.7715 is the net present worth factor of one per period for 9 years at 15 per cent interest rate;
2 0.2472 is the net present worth factor of one at year 10 for a 15 per cent interest rate.

Table 7.6 DCF analysis of a hire rate

Year	*A*	*B*	*C*	*D*	*A + B + C + D = E*	*Revenue, F*	*E + F*
	Capital cost (£)	*Resale value (£)*	*Operating costs (£)*	*Ownership Costs (£)*	*Cash out (£)*	*Cash in (£)*	*Total (£)*
0	−46,000		−0	−4,300	−50,300	0	−50,300
1			−5,000	−4,300	−9,300	x	$x-9{,}300$
2			−5,000	−4,300	−9,300	x	$x-9{,}300$
3			−5,000	−4,300	−9,300	x	$x-9{,}300$
4			−5,000	−4,300	−9,300	x	$x-9{,}300$
5			−5,000	−4,300	−9,300	x	$x-9{,}300$
6			−5,000	−4,300	−9,300	x	$x-9{,}300$
7			−5,000	−4,300	−9,300	x	$x-9{,}300$
8			−5,000	−4,300	−9,300	x	$x-9{,}300$
9			−5,000	−4,300	−9,300	x	$x-9{,}300$
10		+4,000	−5,000	0	−1,000	x	$x-1{,}000$

To return 15 per cent on the investment over 10 years, the total cash flows reduced to net present worth at time zero must equate to:

$$0 = -50{,}300 + (x - 9{,}300) \times 4.7715 + (x - 1{,}000) \times 0.24718)$$
$$0 = -51{,}300 + 4.7715x - 25{,}289 + 0.2472x - 247$$
$$5.02x = 94{,}922$$
$$x = £18{,}913$$

∴ Hire charge = £18,913/2,000 = £9.45 per hour

It can be seen that the DCF method automatically accounts for the depreciation over the life of the asset.

The effect of inflation

The two methods outlined above have not considered the effects of inflation on the value of the investment, the consequences of which are emphasised by the following example based on an assumed 10 per cent per annum rate of inflation.

Example 7.4

Purchase price £46,000
Resale value after 10 years £4,000
£42,000

Annual depreciation $\dfrac{£42{,}000}{10} = £4{,}200$

Inflation at 10 per cent per year
Purchase price after 10 years £4,600 × 2.60 = £119,646
Resale value after 10 years £4,000 × 2.60 = £10,400
£109,246

Average annual depreciation $= \dfrac{£109{,}246}{10} = £10{,}924$
The shortfall = £67,246

Depreciation is the most vulnerable element in the hire rate calculation, because other items are the result of obvious cost movements, that is, materials, wages, interest rates. The hire rate should therefore be revised frequently, to keep up with inflation. In fact, it may even be necessary to provide for 'backlog' depreciation to allow for the period prior to an underestimated price rise. Assessments can be made from special price indices for construction plant, or, alternatively, the small operator replacement prices should be obtained from plant dealers and manufacturers. Inflation produces other severe effects by causing trading profits to be overstated. As a result corporation tax is paid on the inflated rather than the real profit.

Adjustment of a hire rate for inflation

If the capital to purchase an item of equipment is either borrowed or alternatively required by hire purchase, the hire rate would simply be adjusted in line with the terms imposed on the loan, as the total capital sum remains fixed irrespective of the level of inflation. Only the level of interest may fluctuate. However, when the item is to be purchased from internal funds, the value of the asset must be periodically revised to keep up with inflation, if the consequences of a shortfall in depreciation provision and in real terms a negative rate of return on capital are to be avoided. The effect of using a 10 per cent inflation price index on the original example is shown below.

Example 7.5

Inflation is 10 per cent per annum over 10 years. The purchase price is £46,000 and the historical resale value is £4,000 (refer to Table 7.7). The depreciation in, say, the first year of inflation is £4,620.

Interest on finance

With inflation at 10 per cent the apparent rate of return must be used in the calculations:

$$(1 + i_a) = (1 + i_r)\ (1 + i_d)$$

Table 7.7 Adjustment of a hire rate for inflation

Year	*Index*	*Replacement price* (£)	*Accumulated historical depreciation* (£)	*Accumulated inflated depreciation* (£)	*Book value* (£)
0	100.0	46,000	0	0	46,000
1	110.0	50,600	4,200	4,620	45,980
2	121.0	55,660	8,400	10,164	45,496
3	133.1	61,226	12,600	16,771	44,455
4	146.4	67,344	16,800	24,595	42,749
5	160.7	73,922	21,000	33,749	40,173
6	176.8	81,328	25,200	44,554	36,774
7	194.5	89,470	29,400	57,183	32,287
8	214.9	98,854	33,600	77,206	21,648
9	236.4	108,744	37,800	89,359	19,385
10	260.1	119,646	42,000	109,242	10,400

where i_a is the apparent rate of return, i_r the real rate of return; i_d the rate of inflation. Therefore,

$$(1 + i_a) = (1 + 0.15)(1 + 0.1) = 1.265$$
$$i_a = 0.265 = 26.5\%$$

Thus, interest on finance, using a capital recovery factor of 26.5 per cent from interest tables,

$$= \frac{(46{,}000 \times 0.2929 \times 10) - 46{,}000}{10} = £8{,}873$$

Other items

Fixed overhead	£4,000
Road tax and licences	£100
Insurance premium	£200
Consumables	£400
Maintenance	£4,600
	£9,300

Multiplying the total figure by the first year index = (9,300 × 110)/100 = £10,230.

Therefore, the hire charge = £ 4,620
£ 8,873
£10,230
£23,723 ÷ 2,000 = £11.86 per hour (or £474.46 per week)

If inflation continued as shown by the indices, the hire rate for year five should be given by:

Depreciation for year 5 = 33,749 − 24,595	= £9,154
Interest on finance	= £8,873
Other items = 9,300 × (160.7/100)	= £15,531
	£33,558, that is, £16.78 per hour

It can be seen that during periods of inflation the hire rate should be revised at least annually. When inflation exceeds about 10 per cent, the revision may need to be at quarterly intervals, depending upon the demands for payment by suppliers of materials, etc. However, this is not always possible when the market for hired equipment is slack and competitors are willing to undercut hire rates.

When inflation exceeds the net rate of interest on borrowed capital (i.e. interest on borrowings is deducted from company profits before corporation tax is paid, which in effect produces a lower rate of interest), a three year replacement cycle for plant is best. Below that rate of inflation a longer cycle is preferable. The cross-over point in the choice of the method of financing is when inflation is equal to the net of tax interest rate on capital. Below this rate of inflation self-finance is more attractive than borrowed money, and vice versa.

Method of calculating hire rates adopted in practice

In practice the method employed to determine hire rates largely depends upon the size of the organisation and the plant manager's experience and knowledge of prevailing 'economic' market conditions.

Small businesses

The small enterprise commonly sets the hire rate in accordance with the market levels, often guided by the figures published regularly by the Hire Association Europe, Scottish Plant Owners Association, Construction Plant-hire Association, etc.

Indeed, most users are well aware of the prevailing market rate for a machine and each hire is generally negotiated at around this figure. The experienced hirer is usually well informed by the availability of machines locally and thus able to negotiate very competitively. The minimum hire rate, especially for the owner–operator, often relates to payments needed for repayment of the loan on the piece of equipment plus a sum to cover running costs and the operators' overheads, the latter being perhaps as minimal as providing the owner's salary.

Medium-to-large companies

Large companies with comprehensive central and regional plant depots carry out detailed analyses of the economics of owning and operating equipment. The hire rates are usually based on collected cost data of the firm's operations and calculated by the principles established above, including DCF techniques.

The effects of economic recession on plant hire

When the volume of work fluctuates markedly, and particularly with declining demand, many owners of equipment try to cut back the size of their fleets in order to survive. In such a situation a previous good maintenance strategy may assist in temporarily sustaining acceptable resale values. Nevertheless an oversupplied market ultimately forces firms to attempt hiring out plant at unrealistic rates, perhaps abetted by manufacturers offering generous cash discounts on new purchases to hold on to market share. As a consequence, machines get 'run into the ground' as the rates of hire are insufficient to cover adequate maintenance and servicing. Some restructuring of the market then inevitably becomes unavoidable.

Part III

Operational management

Chapter 8

Insurance and licensing legalities

Insurance

Off-highway plant and equipment organisations and equipment hire companies require a variety of insurance policies and cover. This is due to the different legal and contractual requirements for insurance and the diversity of policies, from those to cover liability to those concerning material damage. In this chapter the main areas of insurance cover pertinent to the off-highway plant and equipment will be discussed.

Requirement for insurance

Compulsory insurance

Insurance cover may be a legal requirement and therefore compulsory. Examples of compulsory insurance are the insurance of vehicles as required by the Road Traffic Act, the insurance against claims from employees as required by the Employers' Liability Act and the insurance against claims from the public as required by the Finance Act.

Contractual agreement

Insurance may be required by a contractual arrangement entered into by the organisation and other party. Examples of insurances required by contractual agreements are the responsibilities of contractors under ICE Conditions of Contract or the JCT Standard Form of Building Contract and the insurance required by certain hiring agreements between equipment hire companies and the hirer.

Risk minimisation

Although it may not be a requirement placed on the organisation either by law or by contract, insurance cover may be a wise precautionary measure to take to minimise risks. The Road Traffic Act only requires a specified minimum insurance cover, but the owner of a vehicle may insure it comprehensively, to reduce risk and ensure adequate compensation in the event of loss or accidental damage. Insurance of buildings and insurance of contents are other examples. Some hire agreements between equipment companies and contractors do not specify insurance but place the responsibility for loss or damage of the equipment with the hirer. In such cases insurance is not a contractual requirement but is clearly a sensible precaution.

Classes of insurance

There are four broad classes of insurance available.

1 *Liability insurances*. As far as equipment organisations are concerned the main types are:

- employers' liability;
- public liability;
- motor insurance, third party;
- liability under contract, such as ICE Conditions of Contract or JCT Standard Form of Contract.

2 *Material damage insurance* which covers such items as:

- insurance of works which may be specified in the contract between the employer (i.e. the promoter of the works) and the contractor;
- insurance of buildings and contents;
- insurance of plant and equipment;
- engineering insurance;
- motor insurance covering accidental damage.

3 *Pecuniary insurances* which cover fidelity risks, credit risks and consequential loss.
4 *Benefit insurance* which covers personal accident.

Employers' liability insurance

Employers' liability insurance is required by the Employers' Liability Act. This Act requires an organisation, the employer or master, to take an insurance policy to cover the employer's liability to employees for bodily injury or disease arising out of and in the course of their employment.

Employers' liability insurance is in addition to the responsibilities placed on the employer by the Health and Safety at Work Act. Two issues of this insurance that are relevant to the off-highway plant and equipment industry are those of labour-only subcontractors and hired-in equipment. Labour-only subcontractors present a problem of definition as to whether the labourers are, for insurance purposes, to be treated as employees. The custom has been to regard them as such for the purposes of employers' liability insurance and to have the insurance extended to cover this.

The situation with regard to hired-in equipment with an operator is that while the owner of the equipment, the hiring company, remains the operator's employer, the contract of hire entered into between the owner and the hirer usually requires the hirer to accept liability for the operator. Although the law requires the owner to protect an employee, the contract between the owner and the hirer gives the owner indemnity against claims by the operator which would normally be regarded as employer's liability. The custom has developed for the hirer to indemnify the owner, but if the hirer is not a company but a private individual excavating for some garage foundations, a different view would prevail and the hirer would not be required to carry this responsibility.

Public liability insurance

The Finance Act requires that payments to subcontractors can be made without deduction of tax only if the subcontractor has an exemption certificate from the Inland Revenue.

One requirement in obtaining an exemption certificate is that the organisation must have an insurance policy covering public liability. This insurance is required to cover claims from the public for bodily injury or disease. No requirement exists to cover claims from the public with regard to property damage, although organisations may insure themselves against this also and usually do. Usually a public liability policy will cover claims from the public for bodily injury and disease and for property damage.

The situation regarding hired-in equipment is similar to that of employers' liability, in that hire agreements between equipment organisations and contractors (usually based on the Contractors' Plant Association Conditions for Hiring Plant) contain a clause requiring the hirer to indemnify the owner against claims from the public for injury or damage to property. The common forms of contract found in the construction industry usually require the contractor to indemnify the employer (i.e. the promoter of the works) against claims from the public.

Contractors' all-risk policy (mainly material damage insurance)

In the construction industry a contractors' all-risk policy is usually entered into jointly by the promoter of the construction work and the contractor, and this policy provides the main protection for the works under construction. Usually the items protected against loss or damage in an all-risk policy are: the permanent and temporary construction works and all the materials connected with the works; plant, equipment, tools and temporary buildings and employee's personal effects if not covered by other insurances.

The policy normally includes cover while materials or other items are in transit to and from the site. It is also possible to arrange inflation protection cover in respect of the additional cost of reconstruction after some mishap. The contractors' all-risks policy can be extended to include consequential loss as well as the material damage aspects listed above. The consequential loss cover would protect the promoter of the construction work against loss of income, such as rent from an incomplete building delayed by some mishap. The full value of construction equipment would normally be insured within a contractor's all-risk policy. The ICE Conditions of Contract require the construction plant and equipment to be insured but the JCT form of contract does not.

Hire agreements between equipment hire companies and contractors usually require the hirer to indemnify the owner of the equipment against loss or damage to the equipment. The cover given in a contractors' all-risk policy would normally include the full value of the equipment and therefore protect the hirer against claims from the owner of the equipment.

Plant and equipment insurance (material damage insurance)

Organisations which own equipment and equipment hire companies will normally ensure that their plant and equipment is insured, even though there may be no legal or contractual requirement. Insurance of equipment can be arranged separately or can be arranged as extensions to other policies, mainly the contractors' all-risk policy or an engineering policy, which is dealt with in the next section, or as part of a commercial vehicle policy, if the equipment is mechanically propelled.

An organisation's equipment policy would be a material damage policy and would cover loss or damage while insured. The same cover could be obtained in a contractors' all-risk policy which also includes the travelling of equipment to and from the site. An engineering policy would normally be taken to protect against breakdown and as a means of acquiring an inspection service by a competent person, but the policy may sometimes be extended to cover loss or damage. Mechanically propelled equipment such as dumpers, excavators, graders, etc., may be insured within an organisation's commercial vehicle policy.

In hire agreements between equipment hire companies (the owner) and the hirer, the responsibility for loss or damage of the hired equipment normally rests with the hirer. The hirer may be responsible for lost revenue after the hire period has expired while the equipment is being repaired or replaced. The hirer may not necessarily be required to insure the equipment but will carry the risk if none is taken out. The Contractors' Plant Association Conditions of Hiring Plant are an example whereby the hirer is required to make good to the owner all loss of or damage to the equipment from whatever cause, and in these circumstances the hirer would be well advised to insure against such claims from the owner. This insurance may be included under the policies described above.

Engineering insurance (material damage insurance, breakdown and inspection)

An engineering insurance policy can be taken to cover breakdown and accidental damage risks to lifting machinery, breakdown risks to electrical and mechanical equipment, and the risk of explosion to boilers and pressure equipment. Trucks, tractors and dumpers, etc., could be included in an engineering policy but would be more likely to be included in a motor insurance policy.

If lifting machinery is taken as the main example of interest, the two main risks are mechanical and/or electrical breakdown and accidental damage. Insurance against mechanical or electrical breakdown would include the breaking or burning out of part of the equipment, arising from a mechanical or electrical defect, causing the equipment to stop and requiring immediate repair or replacement. Frost damage would also be covered, but the main exclusion to such policies would be 'wear and tear'. Excavators can be insured under the same policies as cranes, and clauses for accidental damage will cover damage to the equipment from extraneous causes. The engineering policy can also include insurance against damage to property and to the goods being lifted.

The important feature of an engineering policy is that it includes an inspection service by a competent person. The insurance companies who specialise in engineering policies employ qualified engineers to undertake inspections. The inspection service can be acquired separately without an insurance policy and can be arranged to meet the requirements of the Health and Safety at Work Act. The competent person provided by the insurance company can certify that the inspections have taken place.

Motor insurance (liability and material damage insurance)

The Road Traffic Act requires certain minimum insurance for motor vehicles used on the road. A motor vehicle used on the roads must be insured to cover death or bodily injury arising out of the use of the vehicle. The liability is up to an unlimited amount and includes emergency treatment fees. The Act also requires that the policy holder must be issued with a certificate of insurance and that the driver involved in an accident must produce the certificate of insurance.

The motor insurances that are available extend the cover beyond the minimum required by the Road Traffic Act. The four main types are (1) comprehensive, (2) third party, fire and theft, (3) third party only and (4) Road Traffic Act only.

The comprehensive policy provides the greatest cover and includes third party liability for death, bodily injury, emergency treatment fees and damage to property, including other vehicles. The comprehensive policy also provides cover for loss or damage to the insured's vehicle, caused by accidental damage, fire or theft.

Third party, fire and theft policies cover third party liability, as in comprehensive policies, and loss or damage to the insured's vehicle due to fire or theft. Accidental damage to the insured's vehicle, which is covered in a comprehensive policy, is excluded.

Third party only policies cover liabilities to third parties as in comprehensive policies, but exclude loss or damage to the insured's own vehicle.

Road Traffic Act only policies meet the minimum requirements of the Road Traffic Act, which are third party liabilities for death or bodily injuries to third parties arising out of use of the vehicle on the road and fees for emergency treatment.

Special vehicles

Special vehicles can be insured under a commercial motor vehicle policy. Such vehicles or items of equipment can be grouped as follows: digging machines; site clearing and levelling equipment; mobile cranes; mobile equipment – for example, compressors, welding and spraying equipment – dumpers and road rollers.

A goods carrying vehicle fitted with lifting equipment for the purposes of loading goods onto itself is treated as a goods carrying vehicle for insurance purposes. With respect to mobile cranes, the crane itself requires inspection services that are included in an engineering policy. An engineering policy would also provide for breakdown, accidental damage and third party liability, but excludes the liability protection required by the Road Traffic Act. Thus, there is a need for both motor insurance and an engineering insurance on, for example, a mobile crane, and there may be a risk of duplication if the insurance cover is not carefully arranged.

If the mobile crane is hired out, the motor insurance cover can be extended to include use while on hire. Dumpers can be included in a motor insurance policy and the cover can be extended to include use by a hirer. Mobile equipment such as excavators and shovels can be insured as part of a motor insurance policy and can include damage to the insured's own equipment. The cover can also be extended to the hirer. Protection against third party liabilities can be included, but the third party risk other than those arising under the Road Traffic Act may be covered better by a public liability policy.

Site clearing and levelling equipment is grouped separately from mobile equipment, as this class of equipment cannot excavate below the level of the wheel base. Third party risks to property in the ground, such as water, drainage and other services, are much reduced and therefore attract different premiums.

Contingent liability

A motor contingent liability policy is a policy that protects the insured against liability to third parties resulting from the use of vehicles on behalf of the insured, over which the insured does not have immediate control.

As an example, suppose that an equipment hire company believes that the hirer's insurance has been extended to indemnify the owner but that the policy, by oversight, has not been extended, or has lapsed. A contingent liability policy would indemnify the owner in these circumstances. Similarly, a hirer may believe that the owner's motor policy extends its protection to the hirer, but if the owner's policy has lapsed, or is not extended to cover the particular circumstances, then the contingent liability policy could provide protection. This policy is important, because, under hire agreements based on the Contractors' Plant Association Conditions, it is the *hirer* who is responsible for third party liabilities such as injury or property damage caused by the plant. Thus, the *hirer* would bear the costs of such claims if the hire company's policy failed to operate.

Summary of insurance and equipment

This review of insurance policies has described the main areas of insurance cover required by the owners and operators of off-highway plant and equipment. It has also described how insurance can be arranged through a variety of different policies, such as employers' liability insurance, public liability insurance, contractors' all-risk insurance, engineering insurance and motor insurance. The dangers to avoid in arranging insurance cover are duplication, which increases costs, and gaps in the cover, which leave the potential liability or loss uninsured. It is unlikely that such duplication or gaps will arise when equipment is owned and used by the same organisation, because all insurances will probably be arranged with the same company or broker and the insurance needs can be arranged in a co-ordinated manner. The main dangers of duplication and gaps arise when equipment is hired out or hired in, and it is therefore necessary to check all the insurance arrangements in these circumstances.

All organisations require employers' liability insurance and most require public liability insurance. A mobile crane would require motor insurance and engineering insurance to cover breakdowns and inspections. A crane used on site would probably be covered in part by the contractors' all-risk policy but would also require engineering insurance to cover breakdowns and to acquire the inspection service. A field excavator used within a site would probably be covered within a contractors' all-risk policy. Thus, the same policy could cover both an excavator owned by the organisation and an excavator which it hired in.

In general, hire agreements place the responsibility for loss and damage to equipment and the liabilities with the hirer.

Licensing

The use of public roads within the UK is controlled by extensive legislation, broadly called the Road Transport Laws, which control the construction and use of vehicles on the public roads. Within this legislation is a system of licensing which controls the use of vehicles and their drivers on the public roads and collects taxes. This is of relevance to off-highway plant and equipment organisations because some plant items (most notably telehandlers and wheeled backhoe loaders) travel both on and off the highway, generally from one site to another or to deliver materials or parts over short distances. The three main licensing systems of interest are as follows:

1 *Vehicle excise licensing*, which requires a licence, for which duty is payable, to be in force for all vehicles used on the public roads, unless exempted.
2 *Driver's licensing*, which requires every driver of a motor vehicle to hold a driving licence. A prescribed driving test is taken before a driving licence is issued.
3 *Operator's licensing*, a system of goods vehicle licensing which exercises control over operators to ensure the proper use and roadworthiness of the vehicles and observance of the drivers' hours law.

In addition to these three licensing systems, there is *public service vehicle licensing*, which controls vehicles, operators and drivers of passenger carrying vehicles, such as buses and taxis.

Vehicle excise licensing

General requirements

A person who uses or keeps on a public road any mechanically propelled vehicle must have an excise licence in force, unless exempted. Excise licences are issued from local vehicle licensing offices and can also be renewed from post offices. A vehicle excise licence can be taken out for 6 or 12 months. Applications for a licence or its renewal must be accompanied by a certificate of insurance and a test certificate if the vehicle is subject to a test procedure. The annual duty for vehicles varies according to the class of vehicle.

Exemptions

Certain vehicles are exempt from paying duty. Exemptions of particular interest include:

1 road construction vehicles used on a road to carry built-in road construction machinery;
2 a vehicle which is to be used exclusively on roads not repairable at the public expense but details of the vehicle must be declared to the licensing authority;
3 subject to approval, a vehicle which uses public roads only for passing from land in the owner's occupation to other land also in the owner's occupation for distances not exceeding six miles in any week.

Mobile plant

Vehicles carrying no load other than built-in plant or machinery are taxed as goods vehicles. The weight of the built-in plant or machinery is deducted from the total weight in calculating the unladen weight of the vehicle for the purposes of assessing the duty payable.

Driver licensing

No person may drive or permit another person to drive a motor vehicle on a road unless that person holds a driving licence granted under the conditions of the Road Traffic Act.

The minimum age for drivers' licences is 16–21, depending on the class of vehicle. In the UK the minimum age is 17 for an agricultural tractor; for a medium-sized goods vehicle the minimum age is 18; and for larger vehicles it is 21. The minimum age for a goods vehicle not exceeding 7.5 metric tonnes is 18 within the EU and 21 for heavier vehicles, although this can be reduced for drivers who have attended goods vehicle training courses and hold certificates of competence. The driver must pass a driving test, conducted by examiners appointed by the Licensing Authority, to obtain a driving licence.

Heavy goods vehicles

A person must not drive nor permit another person to drive a vehicle classed as a heavy goods vehicle (HGV) unless the driver holds an HGV licence authorising the person to drive vehicles of that class. An HGV is defined as a large goods vehicle which is constructed or adapted to carry goods, the permissible maximum weight of which exceeds 7.5 tonnes, or an articulated goods vehicle. The maximum weight of the goods vehicle or trailer usually will be stated as the maximum gross weight on the Department of the Environment plate, if fitted, or the manufacturer's plate.

Drivers of certain classes or types of vehicle are exempted from requiring an HGV licence. Among the exemptions that are relevant are:

1 track laying vehicles;
2 road rollers;
3 road construction vehicles used or kept on the road solely for the conveyance of built-in road construction machinery;
4 engineering plans;
5 works trucks;
6 industrial tractors;
7 digging machines.

To obtain an HGV licence, a driver must pass a test conducted by a Department of the Environment examiner.

Operators' licensing

General requirements

The Operators' Licensing system is a system of goods vehicle licensing introduced by the Transport Acts and the Road Traffic Act or EU Directives. All goods vehicles exceeding 3.5 tonnes gross plated weight need to be covered by an operator's licence. Exemptions from an operator's licence include the following:

1 agricultural machinery and trailers;
2 dual-purpose vehicles and trailers;
3 vehicles and new trailers using the road for less than six miles per week while moving between private premises;
4 passenger-carrying vehicles and trailers;
5 a trailer incidentally used in construction;
6 road rollers and trailers;
7 vehicles with special fixed equipment;
8 local authority special vehicles for road cleaning and gritting;
9 vehicles used for weighing vehicles or maintaining weighbridges;
10 water, electricity, gas or telephone emergency vehicles.

Vehicles which must be used under an operator's licence include all goods vehicles belonging to the licence holder or in possession by virtue of a hire purchase agreement, hire or loan – that is, all vehicles used by the licence holder, not simply owned by the holder. An operator's licence is required for each operating centre – that is, yard or base for vehicles, in different Traffic Areas controlled by Licensing Authorities. Only one licence is required per Traffic Area, irrespective of the number of operating centres in the Traffic Area.

An application for an operator's licence is made to the Licensing Authority. Application may be made for additional vehicles not yet acquired and, if authorised, this simplifies the procedures of adding additional vehicles to the operator's fleet. The information the Licensing Authority will require includes:

1 the use to which the vehicle will be put;
2 arrangements for ensuring that drivers keep within permitted hours of work and keep proper records;
3 vehicle maintenance facilities;
4 details of any past activities in operating vehicles for trade purposes by the applicant(s);
5 convictions relating to operating vehicles during the preceding five years by the applicant(s);
6 financial resources of the applicant(s);
7 the names of company directors and officers of the applicant company and any parent company or the names of any partners in a partnership.

The Licensing Authority, in deciding whether to grant a licence, will consider:

1 whether the operator is a fit person to hold a licence, bearing in mind past convictions relating to the roadworthiness of the operator's vehicles;
2 arrangements for ensuring that the law relating to drivers' hours and records will be complied with;
3 facilities for satisfactory maintenance;
4 arrangements for checking that vehicles are not overloaded;
5 the suitability of the proposed operating centre;
6 the financial resources for the proper operation of the business;
7 the professional competence of the operator or the operator's manager.

In deciding whether to grant a licence, the Licensing Authority may hold a public enquiry or take objections from a specified group of interested parties, including the police, the local authorities, certain trade associations and certain trade unions. A licence would normally be granted for five years, and additional vehicles may be applied for at the time of licence application or individually when acquired. Arrangements also exist for permanent one-for-one substitution of vehicles.

The Licensing Authority has disciplinary powers to curtail or revoke the licence if a material change in circumstances occurs, or for certain offences relating to roadworthiness of vehicles, drivers' hours and records or plating and testing.

An annual fee is charged for each vehicle specified on the licence and a separate identity disc is issued for each vehicle on payment of the fee. The disc must be displayed on the windscreen of the vehicle.

Maintenance

An applicant for an operator's licence must satisfy the Licensing Authority that the maintenance facilities are such that the vehicles are kept in a safe and roadworthy condition at all times. Licensing Authorities expect more than the minimum routine maintenance specified by the manufacturers and more than the daily running checks made by drivers: they look for a convincing system of inspection and fixed-time-to maintenance. A satisfactory fixed-time-to maintenance inspection system requires:

1 competent staff capable of recognising the significance of defects;
2 a system of recording inspections, detailing what was inspected and the action taken, such as the remedial work done and who undertook the remedial work;
3 adequate facilities for such inspections, including means of under-vehicle inspection;
4 a schedule of inspections the frequency of which is chosen to match the workload and work type of the vehicles;
5 a drivers' reporting system whereby the driver can report vehicle defects.

The Licensing Authority requires that maintenance reports be kept for a minimum of 15 months. Even if operators contract out inspections and maintenance to service companies, they remain the users of the vehicle and are held responsible to the Licensing Authority for the condition of their vehicles.

Drivers' hours

Drivers of goods vehicles are generally subject to EU hours law, controlled by EU Regulation, but there are numerous exemptions.

Drivers' records

EU Regulation requires a tachograph to be fitted and used for goods vehicles registered in a member state. The driver is responsible for returning the tachograph record sheets to the employer and producing records for the current week plus the last day of the previous week.

When tachographs are used, the keeping of a record book is not required. If a tachograph is not in use, a driver must be issued with a record book for the purpose of recording the daily and weekly working records, which are returned to the employer.

Plating and testing

Heavy goods vehicles require a first examination at a Department of the Environment goods vehicle testing station not more than 12 months after first being registered. The first examination includes assessing the vehicle's axle and gross weights. These weights are recorded on a plate, which must be prominently displayed in the cab of the vehicle: hence, these weights are known as the 'plated weights'. The assessment of the weights is followed by a test of roadworthiness, and, if satisfactory, the vehicle is issued with a plating certificate and a roadworthiness test certificate. The roadworthiness test must be repeated every 12 months.

Light goods vehicles are subject to a first roadworthiness test after three years and every 12 months thereafter.

Chapter 9

Equipment maintenance

Introduction

Off-highway plant and equipment must operate efficiently and with minimum unscheduled stoppages, if desirable levels of productivity are to be achieved. Realising this objective will cost the organisation in terms of maintenance works required to sustain production. However, ensuring maximum plant reliability can only be achieved via a carefully planned maintenance strategy. The strategy must coincide with production requirements and schedules so that it causes minimum stoppage and loss of production. It should therefore consider the overall business situation and not only embrace pure maintenance practice but also the wider economic 'cost implications' in the context of the organisation's broader portfolio. The maintenance programme will normally be bespoke and subject to continual monitoring and improvement and will be based upon both practice and experience. Once the maintenance framework has been developed, management must ensure that works conducted conform to policy. The organisation is not selling plant maintenance. Rather plant maintenance is required for its contribution to the overall function of the business.

Maintenance for hired plant: an essential requirement

In today's competitive market environment, many organisations have sold their plant holdings in favour of hiring in plant requirements. It may seem therefore, that such an organisation does not need to be concerned with plant maintenance. However, this view is questionable for two reasons.

Plant hired without an operator

Most plant hire companies will hire out a machine without a plant operator (unless hiring from a specialist hire contractor). Yet plant operators have an intimate relationship with the equipment operated, such that, regular maintenance must be conducted on a daily and weekly basis if the machine is to remain in sound working order. If the operator chosen to operate the plant item is neither trained nor experienced enough in this respect, then breakdown and consequential lost production will almost certainly ensue.

Plant hired with an operator

For plant hired with an operator, plant maintenance is essential. Consider the example of a quantity surveyor reviewing tender prices submitted by plant hire companies (who supply their own trained plant operators) for a forthcoming civil engineering project. Under pressure to keep costs as low as possible, the quantity surveyor announces that the company with the lowest tender bid wins the contract. Unknown to the quantity surveyor, the plant hire company has attained a low tender bid partly through cuts to plant maintenance. Consequently, the plant has a high breakdown frequency which results in lost production for the contractor. The solution to this problem would be to draw up a selective list of reputable plant hire contractors rather than operate an open tender process. Ultimately, a select list will consist of those plant hire companies who maintain their equipment regularly which is reflected in a higher hire cost per hour. This may seem illogical but the benefit here lies in the fact that the higher rate will be offset by the savings made through optimum utilisation. One cannot underestimate the massive disruption that plant breakdown can cause to site productivity. This concept is difficult to comprehend from the client's viewpoint but when considered in terms of, for example, project overrun on time and subsequent loss of revenue, then the need for reliable plant becomes understandable.

Hired or owned, breakdown of a plant item may not cause any immediate financial loss to an organisation directly. However, repeated breakdown will almost certainly result in: delay to contract; possible loss of client goodwill; and loss of company reputation. Combined, these in-turn influence future business. Whilst it would be difficult to assign a cash value to future contracts, one can imagine that lost earnings could potentially be huge.

Maintenance strategies

Maintenance can be defined as 'that which either, retains mechanical plant and equipment in a safe operationally efficient condition; or where plant items have broken down, restores them to safe operational status' (Edwards *et al.*, 1998). By its nature, maintenance involves delays, standing time and lost production. The principal aim of a maintenance department, therefore, is to provide an effective service which attains high availability of plant at the lowest achievable cost. Where possible, costly downtime should be avoided through periodic servicing. Servicing seeks to retain plant in an available state by preventing faults from occurring.

Advanced manufacturing organisations have for a long time recognised the high costs associated with plant and equipment breakdown, typically including inspection, repair and equipment downtime. Breaches of health and safety can also incur financial costs should breakdown cause injury. In their search for improvements to maintenance practice, plant managers within manufacturing industries have developed various strategies for determining when maintenance should be conducted (refer to Figure 9.1). As a consequence of the increased performance of plant and equipment within manufacturing companies, these maintenance practices have subsequently been adopted by a large section of the off-highway plant and equipment industry. This chapter aims to define these strategies and perhaps more importantly, identify their potential benefits.

Fixed-time-to maintenance

Fixed-time-to maintenance (FTTM) requires the implementation of planned regular procedures directed towards ensuring the efficient use of equipment by reducing the incidence of breakdown. FTTM is therefore not a specific type of maintenance but rather the application of maintenance in a predetermined scientific manner. By definition, FTTM is work organised and conducted with forethought. It will include controls to ensure correct and timely

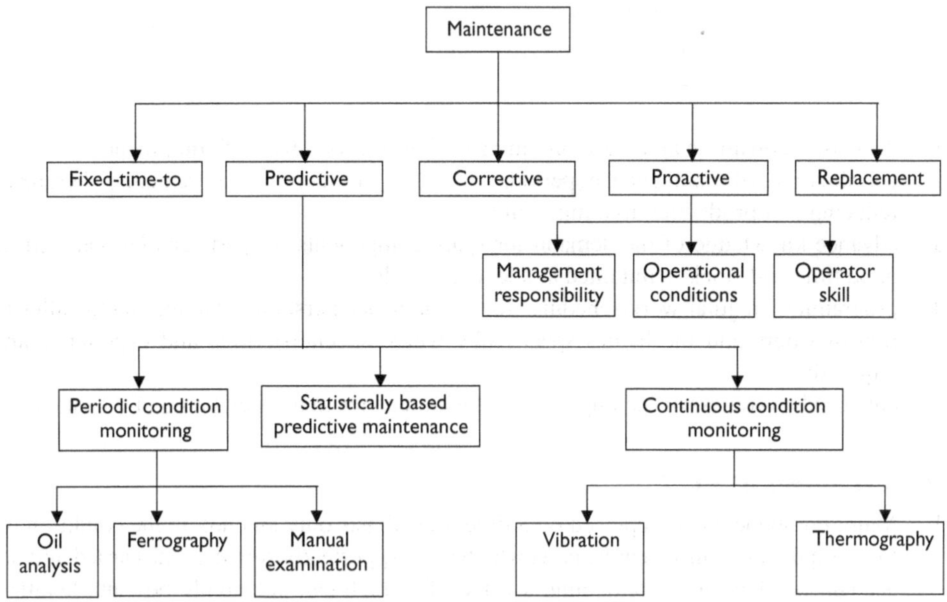

Figure 9.1 Maintenance strategies.

maintenance, and should involve record keeping to ensure that future maintenance is conducted at the appropriate time. FTTM is often called scheduled maintenance and erroneously planned preventative maintenance, since total protection against breakdown is rarely achieved. Put another way, FTTM can only reduce the risk of breakdown and not prevent it; because most failure modes are random in occurrence.

Having established a management framework for maintaining plant and equipment using FTTM, an organisation is well advised to consider the direct activities of maintenance staff. Maintenance actions required in FTTM consist primarily of five key elements:

1 *Periodic servicing*: This will include oil changes, greasing/lubrication, general cleaning, cleaning electrical contacts, adjusting and calibrating component parts, etc.
2 *Component replacement*: Minor components such as oil filters will require periodic replacement within a working period based upon the expected duties and conditions.
3 *Minor repairs*: Minor repairs may be conducted on the machine by either the operator or service engineer and occur as a result of wear and tear. Such repairs may include welding to the structural fabric of the equipment and welding to ground engagement gear, for example, buckets.
4 *Periodic overhaul*: This involves replacement of major components on a time usage basis. For example, a 200 tonne-plus tracked hydraulic excavator has inherent guaranteed service life (by the plant manufacturer) on its components. For example, the guaranteed service life of a slew pump is 7000 h.
5 *Record keeping*: Recording the previous points is essential to ensure that stock and financial control can be monitored and corrected appropriately. Good records also ensure that the service engineer is reminded as to when the next service or component replacement is required.

Advantages of FTTM

The ultimate aim of FTTM is to increase the availability of the plant item, thus increasing production output and hence reducing unit costs of production. Merits of FTTM include:

1 decreased downtime and therefore, improved utilisation levels of equipment;
2 coordination of maintenance periods with site production requirements, therefore, reducing lost production to a minimum;
3 advance knowledge of the demand for spare components, so parts can be obtained in good time and stock maintained at adequate levels;
4 programmed regular work schedules for maintenance personnel, facilitating the allocation of fitters and mechanics specifically where specialist skills and experience are required;
5 enhanced awareness of the importance of maintenance to the well being of the business.

Disadvantages of FTTM

1 Major components are replaced on a time usage basis only and not on the condition of the component. Consequently, parts which may be perfectly sound are discarded and no account is taken of the remaining service life which can potentially be quite lengthy. Because parts are replaced more frequently, the cost of maintenance rises exponentially.
2 Whilst the technique guards against breakdown, it cannot predict it. Hence, when a breakdown occurs costly maintenance ensues.
3 This is a very comprehensive service and can be operated only by those firms with extensive holdings, where 'downtime' can be avoided by substituting other machines and where workshop facilities are available.

Predictive maintenance

Various mechanical signs, conditions or other indications precede 99 per cent of all mechanical failures. With this in mind, non-destructive techniques have been developed to determine incipient failure. Indeed, equipment diagnostic techniques are vital to the successful maintenance and operation of off-highway plant. Condition based predictive maintenance (CBPM) is founded upon the principle that wear is responsible for innumerable mechanical breakdowns and even where not directly responsible, some element of the phenomena is usually present. Wear is a gradual process which affects components at varying rates. The process will not normally cause sudden mechanical failure, but rather, wear is preceded by changes in a machine's sensible behaviour. CBPM focuses on the examination of these wear processes on mechanical components and is normally conducted by mechanical engineering experts (often plant manufacturers offer such services to the plant owner). The specific conditions measured will vary dependant upon machine type. However, most measurements provide quantitative data on some or all of the following:

1 temperature – bearings, lubricating oil and electric motors;
2 vibration – rotating machinery such as pumps and compressors;
3 pressure – fluid distribution systems, for example, oil supply;
4 electric current – electric motors and pumps.

A useful analogy for understanding CBPM is to compare mechanical plant failure to serious illness in the human body. For instance, the enhanced probability of a patient making a full recovery following a sudden heart attack, requires the application of immediate paramedic aid. This is analogous to emergency repair and maintenance being performed on broken-down plant items. However, such reactive diagnosis may lead to 'on the spot' prescribed actions which may only prematurely postpone death. In the case of plant and equipment, this would be equivalent to irreparable damage! The important lesson here, is that by conducting regular health checks, the signs and symptoms of incipient breakdown can be identified. CBPM can be further classified as being periodic, continuous and statistical.

Periodic condition monitoring

As its name suggests, periodic condition monitoring entails monitoring the condition of mechanical plant at regular, predetermined intervals and usually coincides with a FTTM program. The time interval between observations must be significantly less than the expected time to failure of component parts, or time to deterioration of lubricants around these components, if unplanned stoppages are to be averted. The principal technique used for the majority of off-highway plant and equipment items is *tribology*. Tribology can be defined as the observation of 'interacting surfaces in relative motion'. The term is derived from the Greek word 'Tribos' meaning rubbing and it translates into the English language as the 'science of rubbing'. Tribology encompasses surfaces sliding, rolling or otherwise interacting with each other. Hence, inevitably, the problems of friction and wear are an integral part of the science.

In order to inhibit the detrimental effects of wear, friction and heat generation within mechanical machinery, lubricants are used. Elementary chemistry demonstrates that lubricants are simply a mixture of long chain hydrocarbons mixed with other speciality chemicals. One important property of hydrocarbon molecules is their immunity to physical attack. For instance, they cannot be destroyed by grinding between gears. However, lubricants can become contaminated in use thus reducing their excellent inherent friction resistant properties. Due to lubricants' role in mechanical machinery, a basic understanding of the key processes involved in their degradation is necessary. Essentially, oil degradation will result from either oxidation or contamination, or a combination of the two.

1 *Oxidation*: This is the process by which a lubricant loses hydrogen thus altering its molecular properties. The intrinsic properties of a lubricant sample should therefore be considered against ideal standards so as to determine its current health.
2 *Fluid contamination*: Other fluids, particularly water, can find their way into lubrication oils. Because these contaminants fail to minimise friction and wear to the same degree as oil, they inevitably accelerate deterioration of the mechanical components when present.
3 *Solid contamination*: Mechanical action may generate excessive stress beyond the elastic limits of metal components and therefore induce fragments to break off component surfaces. This deterioration process may then be further exacerbated as these microscopic fragments (now present in the lubricant) accelerate mechanical stresses and thereby in turn increase the rate of component erosion. Intuitively, it makes sense that contamination free lubricants provide better service to an item of mechanical plant.

In the context of CBPM, tribology analysis is the generic term embracing a range of techniques which scrutinise oil condition. Each technique endeavours to examine specific problems which may be exposed by examination of the oil sample, such as wear particles and contamination. In many applications, lubricants serve to wash away loose particles generated via mechanical action. Resultant study of the particle shape, size and composition residing in an oil sample, provides direct insight into mechanical plant's inner condition. Samples of the oil (extracted from the various compartments of a machine) are normally taken by the operator but specialist chemical engineers within the manufacturer (e.g. Caterpillar) or petrochemical companies (e.g. Castrol) often conduct the analysis for the plant owner.

Some degree of wear out is inevitable, but the charting of tribology analysis enables progressive wear out trends to be estimated, together with the types of wear (i.e. rubbing wear, sliding wear or rolling fatigue). Incipient wear out can then be immediately detected so as to pre-empt breakdown by applying *just in time* action. Principal tribology analysis methods are oil analysis and ferrography, each of which is now discussed in some further detail.

Traditional oil analysis

Traditional oil analysis measures viscosity, water content, acid number and other physical parameters against benchmark standards. However, in recent years, use of the technique has been supplemented with spectroscopy. This is a technique which serves to identify contaminants by measuring their characteristic electromagnetic wavelengths when introduced to a high energy source such as plasma. The limitation of spectrographic techniques relates to their inability to give quantitative analysis of particles over 10 μm in size. For larger particles, ferrographic techniques must be used. Ironically, this disadvantage is also the major strength of spectroscopy because measurement of small particles (1–5 μm) is of greatest value in determining machine running condition. Large and medium particles (0.1–0.5 mm and 10–50 μm, respectively) tend to exit oil circulation either via filtration or settlement. Conversely, small particles which are less prone to settlement remain suspended within the system thus their measurement provides true indication of machine condition.

Ferrography

Ferrography, as a technology, is particularly useful for measuring lubricant quality in one of the most fundamental of all engineering mechanisms – the gear. Increasing performance of today's gears (and other components) compounds the need to monitor the quality of lubricants, so as to ensure prolonged life and efficient running of machinery. Measurement of wear particles using ferrographic techniques is made by slowly pumping the lubricant across a glass slide exposed to regions of successively greater magnetic field strength. Where non-magnetic particles require analysis then the technique may be enhanced further by the judicious use of calibrated membrane filters. This initial stage provides quantitative analysis by grading the particles present in the sample from small to large, to give the wear particle concentration (WPC) ratio. This is useful for observing trends in WPC, particularly large particles, whose concentration will increase significantly under abnormal wear. The second stage of testing, comprises qualitative analysis of particles which are examined under a microscope. This enables shape, colour and thus the origins of wear to be determined. More interestingly, from the predictive stance, particle analysis generates information relating to the rate of wear progression. The principal types of wear can be categorised as one of the following:

1 *Scuffing*: High loads or speeds generate excessive heat which breaks down the lubricant between mating teeth. Adhesion ensues, causing roughening of the wear surfaces and a subsequent increase in wear rate. Due to the nature of scuffing, particles present share the common properties of a rough surface and jagged circumference. In addition, the high temperatures generated cause oxidation which manifests itself as a bluing on the particles.
2 *Severe sliding or overload wear*: Excessive stresses induced by high speeds or loads will result in severe sliding wear, causing large particles to break off components. Further elevation in stresses leads to a second transition being reached; where surfaces breakdown and a catastrophic wear rate ensues.
3 *Fatigue*: Fatigue occurs as a result of excessive tensile stresses upon the mechanical surface, which initiates fatigue cracks. Ultimately, total failure will occur.

Manual examination

The benefits of a relatively simple manual inspection of plant by an experienced operator, should never be underestimated. Such inspections can yield vital information on machine condition with minimum intrusion or cost. In essence, the four senses of sight, sound, touch and smell are used to determine the condition of machinery. Hence, a leaking hydraulic pipe can be detected visually, a fault in an exhaust system can be heard and smell or touch can detect an overheated electric pump or motor. Together with data collection for future analysis, visual inspection can provide vital condition monitoring information. The principal benefit of manual examination is realised when used in conjunction with CBPM analysis. During the processing of CBPM data mistakes can be made since neither computers nor data processors are ever totally infallible. Unusual observations which do not fit previous patterns can therefore be checked for validity and the discrepancy either rejected or confirmed.

However, human inspection has three principal disadvantages. First, information generated tends to be qualitative and thereby subjective. Second, it is heavily dependant upon the skill and judgement of the individual operative. Finally, one should consider the fact that in most cases the technique can only detect component failure once a fault has already occurred, at which point some damage has been sustained. Therefore, manual examination is largely limited (although not exclusively) to being predominantly a 'post diagnostic failure' method of condition based monitoring.

Continuous condition monitoring

Continuous condition monitoring (CCM) involves monitoring some condition or variable associated with an item of equipment, on a continuous basis. As such, condition monitoring equipment is an integral part of the machinery. Due to the 'continuous' emphasis of this method of CBPM, it can be quite expensive to initiate and perform. Consequently, it is normally confined to equipment which is in constant operation, such as that found in manufacturing, and where the cost of lost production in the event of breakdown, is extremely high. For off-highway plant and equipment, the use of CCM is currently scarce and largely confined to larger (and more expensive) plant items operating within the coal mining sector.

For example Liebherr, Komatsu (Mannesman Demag) and Caterpillar all utilise CCM technology. Liebherr are however particularly advanced in this field since the development of an integrated onboard monitoring system that is linked via global satellite positioning systems to remote PCs owned by both the manufacturer and the owner. This technology is set to change the way with which plant and equipment items are maintained albeit, it may be many years before the technology is fully adopted within industry.

The technology that is integrated into the continuous monitoring system can consist of either vibration monitoring, thermography or both.

Vibration monitoring

Due to minor imperfections, rotational or reciprocating parts found within mechanical plant (such as gear boxes, shafts and engines), produce vibration frequencies at various signature amplitudes. The vibration occurs naturally to some extent, as a result of the forces and mobility created in components of a machine when undergoing normal operation, such that; *force* × *mobility* = *vibration*. During use, moving parts wear or deflect at an increasing rate as clearances between components increase and rotating parts become further out of balance. By careful study of vibration levels, an assessment of the condition of individual component parts can be made.

Vibration monitoring is to some extent a compromise, because in an ideal situation it is the forces themselves (i.e. those causing the vibrations) which require measurement. This is practically impossible, so a compromise is sought by measuring the vibration, which has an indirect link to the force through a factor of mobility. Caution is necessary because a high vibration level does not automatically indicate a faulty component. Such vibration may result from a high level of mobility within that component. Periodic measurements are therefore taken on the basis that the vibration amplitude will remain constant unless a change occurs in the operating dynamics, that is, a defect occurs. Once the wear process has begun, extra deterioration of the component will increase the rate of wear. Consequently, a new component will run smoothly until wear becomes noticeable, at which point the component will deteriorate progressively and manifest itself as ever increasing levels of vibration.

In practice, baselines of vibration characteristics are mapped for new components to serve as benchmarks against which to compare future vibration signatures (and thereby facilitate machine diagnosis). Regular monitoring of vibration levels and comparison of these levels to benchmark standards will alert plant managers to wear trends. This provides an advanced 'lead time' warning of component failure.

The recent rapid development of personal computers and business software has produced specific vibration control systems which further enhance the powers of spectral analysis. Modern packages now offer fast and effective data collection, data measurement, analysis and diagnosis.

Thermography

Thermography is the technology of measuring components' infrared energy emissions using thermographic instrumentation. Thermal imaging was originally pioneered in the steel industry for monitoring the condition of blast furnace linings, stoves, steel mixing vessels and electrical machines and equipment. In a maintenance context, deviations in temperature conditions (components being hotter or colder than specified limits) are indicative of pending problems. To the plant engineer, thermography is particularly well suited for sensing excessive heat generated by friction in faulty rotating components of mechanical equipment. An inoperative cooling valve will induce an increase in machinery temperature, which in turn stimulates a catalogue of chain reactions, for example, leaking gaskets, thermal viscosity, breakdown of lubricant protective qualities and engine seizure.

Thermographic techniques are sensitive to variations in ambient conditions such as the amount of air borne particles, so compensation for variations in measurement must be taken into account when analysing thermal data. However, the principal benefit of using infrared techniques in mechanical inspections, lies in their unique ability to instantaneously pinpoint problematic areas and thus reduce labour maintenance costs. Moreover, advances in microtechnology have inspired ever more compact 'hand held' equipment which is invaluable to plant managers operating on remote sites.

Statistically based predictive maintenance

Statistically based predictive maintenance (SBPM) involves meticulously recording all stoppages in plant items and components. The statistical data generated then facilitates development of multivariate statistical models for predicting failure and thus, enables maintenance measures to be undertaken via a FTTM maintenance policy. Due to the immense variability associated with off-highway plant (i.e. a multitude of ground conditions encountered and numerous operators of various ability), the use of SBPM has historically proven difficult to administer. However, the use of advanced statistical software packages (such as Statistics for the Social Scientist (SPSS)) has automated the process of analysis and facilitated wider and easier application throughout industry.

Advantages of predictive maintenance

Predictive maintenance techniques are claimed to have several key benefits over fixed-time-to maintenance strategies. These benefits include:

1 *Reduced capital lockup*: Plant holdings that conduct regular 'in-house' maintenance will require a well stocked parts and spares department, to replenish and renew components as part of a regular fixed-time-to maintenance strategy. The benefit provided by predictive maintenance techniques is that plant managers can better forecast the remaining life of components (in use). Therefore, replacement parts can be purchased in good time (rather than held in stock), reducing the need for large (costly) store supplies.
2 *Reduced incidence of breakdown through forecasting*: Breakdown is a double-edged sword. During breakdown one first loses the ability to produce, whilst the costs of overheads and other supporting resources (e.g. labour, stores, etc.) still continue to accrue. To accentuate matters, in an attempt to maintain productivity rates short-term hired machines may have to be employed on site which means further direct hire costs and costs of administration. Total maintenance costs can therefore be significant. Herein lies the principal benefit of predictive maintenance, that is, the ability to monitor the mechanical health of equipment components and based upon this analysis, conduct only the necessary maintenance required to maintain near 100 per cent operational reliability.
3 *Reduced direct maintenance costs*: Consider once more the fixed-time-to stance which replaces component parts on a time-usage basis. An example was given earlier of a slew

pump (on a tracked hydraulic excavator) which had an expected life of around 7,000 h. Under the fixed-time-to approach to maintenance, the slew pump would be removed and replaced at around 7,000 h with new, in order to prevent breakdown. No account is taken of the remaining life of the component, except perhaps its residual and therefore second-hand value. Yet that component could have had a remaining service life of many thousand hours. This is one of the main criticisms of fixed-time-to maintenance and therefore, highlights the benefits of a predictive approach.

Disadvantages of predictive maintenance

The advantages of predictive maintenance far outweigh the disadvantages. However, there are two disadvantages which should be considered. These are:

1 *Failure to determine root causes of maintenance*: One could argue that although predictive maintenance can identify the symptoms of mechanical wear and hence, component failure, by the time the imbalance in mechanical physiology has been identified, irreparable damage has already occurred. This point is particularly relevant to organisations who use periodic condition monitoring on infrequent and irregular intervals because the wear trend cannot be found.
2 *Management interpretation of results*: One of the major criticisms of predictive maintenance lies not with the technique but with the limited interpretation of its results once attained. In the case of used oil analysis (a common predictive technique), only recommendations of analysis (given by the analysis specialist) are usually acted upon. However, the analysis reveals far more than just a recommendation, since it will enable the performance of the plant item over time to be determined. The rate of wear of components; the factors influencing wear; and optimum strategies for prolonging component life are just several examples of the type of information which can be extracted from results. The problem is further compounded by the widespread practice of disposing of used oil analysis result sheets soon after use. Consequently, the ability to build a comprehensive history of machine operating conditions is lost.

Corrective maintenance

Corrective maintenance is performed only when an unpredictable fault occurs. The ethos of corrective maintenance is to restore equipment to a safe and operationally efficient condition in as short a period as possible. Normally, diagnosis of the fault or failure is followed by implementation of corrective action which may include:

- adjusting, replacing or repairing the parts, components and subsystems which caused the unscheduled failure;
- checking the plant is operational again;
- cleaning the machine and logging the restorative action, action time and cost.

Corrective maintenance is valid maintenance practice since no matter how technologically advanced the strategy adopted, plant managers may still be faced with the odd breakdown which can neither be predicted nor actions taken to prevent its occurrence. Thankfully, such occurrences are rare amongst modern machinery and mainly limited to faulty components early on in a machine's life.

Replacement

Most items of plant and equipment have a life exceeding the point at which a major overhaul is required, and the question of replacement should therefore arise only when the costs of maintenance exceed the benefits of operating the item. However, the state of the second-hand market can fluctuate over the short term to provide a profitable opportunity for selling the machine before its planned replacement period. Alternatively, superior equipment may become available, to outweigh the advantages of holding outdated machinery.

Proactive maintenance

A relatively recent area of maintenance philosophy is root cause failure analysis. This takes a proactive stance to maintenance by, as its name suggests, examining the source of breakdown. The philosophy of proactive (as opposed to reactive) maintenance concentrates on commissioning corrective actions to the underlying factors which initiate fault. From the engineer's standpoint, the root causes of maintenance relate to the factors which alter a lubricant's protective properties, for example, soil, fuel and water contamination. However, from a management perspective, the true root causes of mechanical fault include inadequate operator training, poor maintenance specification/regime, inadequate plant maintenance policy and adverse operational conditions. This is because internal component contamination arises as a result of these factors. An understanding of the long-term benefits of controlling the root causes of machine failure must be sought. Each of these root causes warrants further discussion.

Operator skill

Technologically advanced industries, such as manufacturing, enjoy the comfort of fully automated machine trains which literally take resources in and push products out. On-board computer aided monitoring ensures that optimum working levels are achieved, whilst labour input is effectively cut to a minimum. Against this backdrop, CBPM is the perfect solution to pre-empt breakdown via fault detection. However, within the off-highway plant and equipment industry, there is an interdependent relationship between machines and their operators. Unlike the technologically advanced machine trains used in manufacturing (many of which have continuous condition monitoring), off-highway plant and equipment is dependant upon operator competency to manually monitor machine condition and take periodic samples for the various machine compartments. Such activity is performed in isolation due to the bespoke nature of off-highway projects, an exception being plant which operates in opencast mining where contractors usually have their own plant department on site. Within general industry, the burden of daily predictive and fixed-time-to maintenance responsibility lies heavily upon the shoulders of the plant operator.

In most cases, plant operators have limited time to perform maintenance on their plant. Instead, the operator may 'work around' maintenance needs in an *ad hoc* way. When a breakdown occurs emergency 'restorative' action is often taken to return the equipment to working order as quickly as possible, as opposed to fixing the fault correctly. Over time and use, the cumulative effect of deferring maintenance and performing inadequate repairs culminates in plant requiring replacement before its expected life, or plant which breaks down with increasing frequency and for longer time periods. Either way, the negative impact upon business is profound.

The operator has an essential role in fixed-time-to maintenance. Oils must be changed at regular intervals, greases applied to grease points, the equipment cleaned and many other minor maintenance tasks conducted. Emphasis on maintenance strategy (i.e. predictive or fixed-time-to) is therefore changing to a more proactive approach. Operator competency is one of the most important factors to be considered since it is the operator who implements maintenance directly. Success or failure of a maintenance strategy will largely depend upon the operator's attitude in this respect. However, management must ensure that the operator is sufficiently motivated, trained and competent to conduct maintenance correctly and in a timely fashion. Lack of routine maintenance scheduling, maintenance materials purchase and poor supervision can all have a de-motivating effect upon the operator.

Operational conditions

Off-highway plant and equipment is employed in a host of extreme weather and site conditions. Environmental factors are particularly prominent in the mineral extraction industries where a high proportion of severe wear problems arise from abrasion. Typically in the UK one can expect to be exposed to a barrage of silicon (dirt) and water, hence seals on equipment must effectively prevent contaminants from entering internal compartments. Two factors can reduce the amount of mechanical wear caused by operational conditions.

Manufacturer design

Perhaps the most significant positive contribution to proactive maintenance is going to be made by plant manufacturers themselves. Improved wear resistant materials, systems, designs and monitoring techniques to minimise the effect of inhospitable conditions (e.g. underground coal mining) will both reduce the need for maintenance and extend the time between maintenance periods.

Appropriate specifications and design

Proactive management matches the most appropriate maintenance strategies to prevailing operational conditions. For example, excavators are expected to work on an endless range of operational conditions varying from dry humid environments to freezing desolate landscapes. Each environment presents new challenges and obstacles that must be first assessed and then guarded against. In a hot, dry atmosphere where under track conditions are abrasive, silicon ingression into fluid filled compartments and increased wear of ground engagement gear (e.g. tracks, buckets, shovels and undercarriage) present major problems. Under such conditions, the standard maintenance regime employed will have to be upgraded considerably to include the use of bucket teeth constructed from hardened steel and perhaps more frequent oil, lubricant, gasket and filter changes. Whilst such may increase FTTM maintenance costs it will also reduce breakdown and associated costs.

A similar scenario exists when wheeled machines operate in the scrap metal (or similar) industry. By nature of the under tyre environment, scrap metal, nails, bolts and other debris litter the surface and are responsible for numerous punctures. The problems here can be rectified through a combination of appropriate machine selection and good housekeeping. If machines do not travel extensively then a solution to puncture failures would be to use a tracked machine. When travel restrictions are not possible, then a viable and practical solution would be to maintain traffic routes such that debris is kept to an absolute minimum. One should always consider the cost of good housekeeping compared to the costs of increased maintenance and reduced production. In either example, these findings provide conclusive evidence that operational conditions significantly influence machine maintenance costs.

Management responsibility

The off-highway plant and equipment sector continuously witnesses changes in maintenance procedures and practice, equipment sophistication and equipment diagnostic procedures. Ultimately, such changes contribute to the continued evolution of the modern plant manager. Senior management must adapt to such technological changes for continued success of the organisation. With advancements in information technology (IT), for example, new breakthroughs will continue to occur at an increased rate. Embracing continued developments in maintenance technology and practice will lead to improved quality control, better plant information handling, lower costs and a more highly skilled operator and line management work force. Unfortunately, maintenance is one area where established management principles are sometimes ignored. Rather, maintenance can be viewed as an encumbrance. Management must give the same attention to plant maintenance as given to more 'traditional' responsibilities such as book keeping, cost control, personnel management and leadership.

In plant management practice, education (or rather, lack of continuous education), forms an impervious barrier preventing many potential benefits of technological change from being embraced. Continuous training is vital for organisations who wish to maintain a competitive edge over rivals since employees can become obsolete within a mere three years of training completion. Consider guidelines for operating and maintaining plant and equipment which are currently widely available to the plant owner. Not all plant owners treat their equipment according to specifications. Problems such as incorrect specification of oil are often encountered. This particular problem would be detected as an imbalance of additive levels by the oil analysis but by the time of detection, irreparable damage may have already occurred.

Management should also consider the wider implications of a maintenance stratagem which fails to maintain historical maintenance data, for example, a complete history of used oil analysis results. Because records are often forsaken for the benefit of space saving, the opportunity to trend plant wear under operating conditions and thereby adjust maintenance policies to optimum levels is ultimately lost. Maintenance management thus fails to progress along the learning curve to enhanced utilisation of equipment. A strategic approach which pays equal emphasis to managerial roles and responsibilities is therefore required. This will involve the development of an efficient plant maintenance policy which embraces:

- an effective programme of service repairs;
- clear guidelines for servicing responsibility;
- an effective spare parts inventory or delivery system;
- a clear method of managing the flow of paperwork.

Senior management may consider the adoption of ISO 9000 'Quality Management and Quality Assurance Standards'. ISO 9000 should not be confused with Quality Control (QC) which refers to the process of controlling quality and the employees who achieve it. QC was popular in production departments but its popularity dwindled as the system was based upon a false premise. That is, quality was external to production and that higher quality was achieved through inspection and consequential acceptance or rejection of poor quality items or practices. Quality management professionals realise today that quality management must be a comprehensive program of continuous improvements to the quality of an organisation's products, processes and services. Hence, ISO 9000 sets standards for evaluating the system of quality management within an organisation *vis-à-vis* setting a standard for a product. Essentially ISO certification allows management to question the quality of maintenance records and moreover, facilitate periodic external evaluation by a certified company (by ISO standard officers) to determine continued compliance.

Since senior management have ultimate control over the wider area of management practice and quality assurance which must support maintenance practice within the organisation, management responsibility is one of the most important factors which denotes the success or failure of maintenance practice.

Management responsibility: controlling fluid contamination

From a practical standpoint, management can prolong the service life of off-highway plant by adopting an effective fluid contamination control mechanism. This should essentially provide a clear action plan with which to prevent internal components from becoming contaminated in the first place. This will consist of the following stages:

1 *Determination of acceptable levels of contamination*: The level of cleanliness required must be based upon machine operating parameters. These will include operational environment and machine sensitivity to contamination.
2 *Attainment of cleanliness targets set*: Installation of a new oil filter is not only expensive but also time consuming. In fact Edwards *et al.* (1998) suggests that it can cost up to 10 times as much to remove a particle once it has entered a machine's fluid, than it costs to keep particles out in the first place. In-house contamination begins as soon as new fluids are purchased, since new oil can be contaminated during transportation. Hence, inspections should be made of new containers to determine broken seals and to check that containers are clean and free from puncture. Moreover, during machine oil and fuel changes the manner in which machines are filled should be monitored. This will involve ensuring that equipment used (e.g. funnels) is clean and free from debris.
3 *Continued conformance to cleanliness standards set*: This will pursue effective contamination control through rigorous enforcement of cleanliness standards. Management will need to set standards and procedures for changing oils, conduct regular monitoring of oils in service and control oil movement from purchase to disposal.

Advantages of proactive maintenance

The principal objective of proactive maintenance is to prolong machine life and thus negate the need to conduct repairs when no failure has occurred. The concept therefore fuses the salient advantages of both predictive and FTTM techniques and leads to enhanced mechanical maintenance efficiency. Adoption of this approach in industry would therefore improve upon the benefits already offered by CBPM techniques. Such benefits include:

1 less downtime, resulting in increased productivity and increased utilisation;
2 reduced spares stock and lower capital lock-up;
3 savings through reduction in frequency of condition monitoring costs;
4 potential reduction in accident occurrence, both on site and during maintenance;
5 reduced manpower to implement maintenance regimes.

Many of these benefits will result in reduced expenditure. However, benefits should be set against new costs associated with changes in existing maintenance management systems to facilitate root cause failure prevention. Obviously, this relationship will depend upon an organisation's previous maintenance expenditure, type of plant employed and the intangible costs associated with breakdown.

Strategy for plant maintenance

The essential aim of all the maintenance policies described is to keep equipment in working order and so increase its productivity. However, the strategy required to achieve this objective demands the implementation of technical and administrative procedures which inevitably incur costs, and for any organisation, depending upon its maintenance efficiency, there is an optimum level of maintenance provision, as shown in Figure 9.2. This implies that at some level the cost of providing the maintenance service will exceed the costs of machine 'downtime'. Thus, not only is it important to install the correct maintenance procedures, but also the costs must be maintained and controlled. These can be considered in two classes – direct and indirect costs.

Direct maintenance costs

The first step is to prepare a maintenance budget based on the needs of the equipment holding and any items to be added during the life of the budget. Reference should be made to historical records of:

- breakdown maintenance labour costs;
- planned maintenance labour costs;
- materials costs and fuel consumption;
- spares costs;
- administrative, technical, equipment and other overhead costs.

The budget provides the basis for monitoring the trend in overall maintenance effectiveness. In order to collect costing information, each item is identified with a cost centre, which

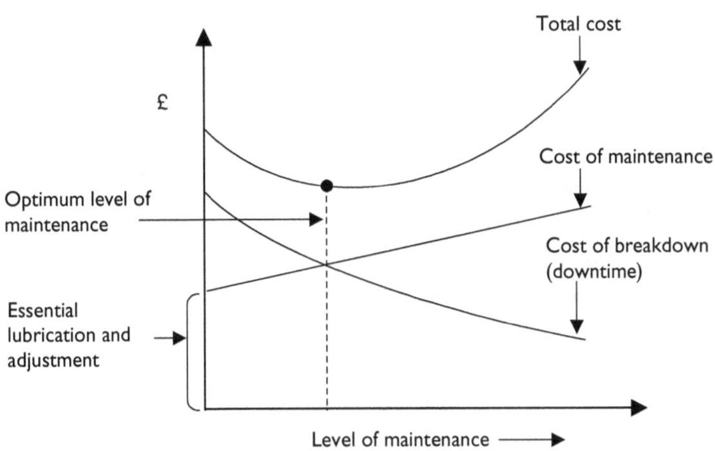

Figure 9.2 Optimum maintenance provision.

might represent a single important piece of equipment or alternatively a group of similar machines. The extent and depth of the information will vary between firms but, as a basis for information flow, a cost and maintenance control system should contain the following features:

1 *A register of all equipment assets*, detailing the type, classification, purchase price, location, life, age, value and condition of each.
2 *A schedule of the exact maintenance tasks* required on each item on a routine basis, together with the extent, details and methods to be employed.
3 *A programme of events* defining the frequency of these maintenance takes.
4 *An effective history record* for each item, to ensure that the maintenance has been performed on schedule and in the correct manner. This will include a weekly return of hours operated or miles travelled and fuel consumed, coupled with a record of maintenance works carried out and the cost, including the materials used and spare parts supplied. The history record card is updated regularly with maintenance information each time a job report card has been completed.
5 *A cost recording system* to facilitate monitoring the effectiveness of the maintenance effort measured against the budget.

The effectiveness of this procedure depends upon a disciplined application of the checking system and should ideally follow the stages indicated in Figure 9.3.

Job report card

The job report card should be completed by the mechanic after work has been carried out, and should contain a statement of the work done, the materials used and the condition of the unit. Subsequently, this information is used to update the history record card (see Figures 9.4 and 9.5).

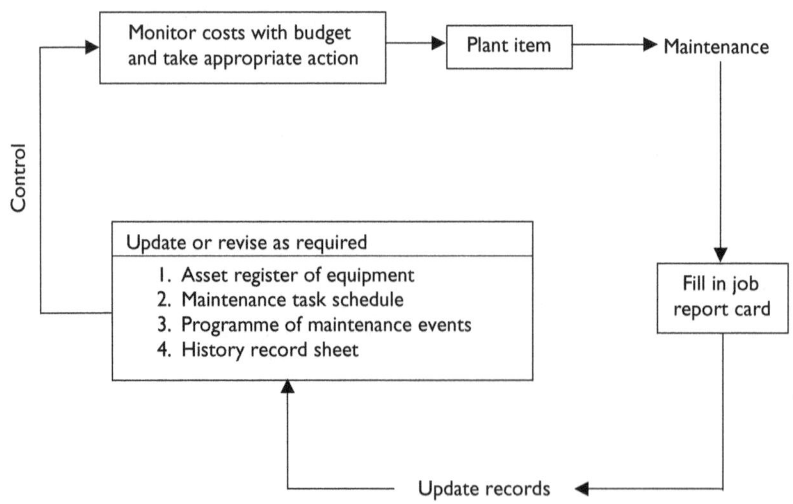

Figure 9.3 Maintenance control cycle.

Job Report Card No.:		**Date:**
Name and details of mechanic:		
Equipment model and make:		Location:
Equipment ID no.:		Hours worked:

Defects: ....

Action taken: ....

Spares/Materials used: ....

Condition/Observations: ....

Remarks: ....

Time taken: ....

Remaining defects: ....

Figure 9.4 A job report card.

Indirect costs

When equipment breaks down, there is a loss of production while waiting for repair and during the repair itself. For the hire firm this may cause a reduction of the budgeted utilisation period and a loss in revenue, but for the owner–contractor the idle time of the workforce must also be considered. It is important to consider these costs when preparing an overall budget for owning and operating the machine, including the cost of maintenance. Clearly, it is essential to decide upon realistic levels of utilisation time over, say, a yearly budget.

Safety inspections

The various safety regulations (see Chapters 8 and 10) require the inspection and testing of equipment. All lifting tackle and most other types of equipment, when first acquired and

History Record Card

Machine ID no.:	Location:	Machine hours:
Inspection no.	Date:	Engineer:

Mechanical fault(s): ...

Description: ...

Remarks: ...

Components used: ...	Time	Cost
Consumables used: ...		

Time at commencement: ...

Time of completion: ...

Figure 9.5 A history record card.

before use, should be tested and examined by a competent person who should certify that the equipment is suitable for use and issue a certificate stating so. In most cases this certificate will be provided by the manufacturers but, if not, an insurance engineer can issue such a certificate. Once the item is in use the regulations also call for periodic testing by a competent person, the time interval depending upon the type of equipment, and a certificate issued by that person.

Because the various regulations stipulate different inspection periods, the recommended procedures to adopt should consist of a combination of self-inspection by the owner or hirer at say one month intervals, and then external examination within the legally stated testing period. This latter inspection could be performed by a qualified engineer from the engineering inspection department of an insurance company. The services of the insurance company, however, do not relieve the owner of the equipment from the legal responsibility for periodic examination. Although most insurance companies have their own methods to try to ensure that the statutory periods are met, they are not legally bound to do this. Also, the insurance engineer having arrived at the premises will not necessarily search for equipment to test or get it ready for examination, and if the items are not to hand, then the inspection will most likely be missed. Clearly, therefore, it is in the interests of the equipment owner to install the proper control procedures for inspections.

Control procedures for testing and examination

Whenever possible, responsibility for arranging testing and examination should be given to a single individual. In this context, the asset register can assist in the time-tabling of inspections for major equipment, but for *small* items, such as lifting tackle, a physical audit may be required, since individual employees and machine operators often hoard items such as slings in tool boxes, lockers and cabs. To minimise this location problem, all equipment should be issued only through the stores under the control of a storekeeper and records maintained at least to the standard demanded by the legal statutes. The law requires that records be kept of the following details for each item:

- an identification mark code or number;
- the date and reference of maker's original test certificate;
- date of commissioning;
- dates of statutory tests, inspections and examinations;
- details of defects and subsequent actions required;
- copies of test certificates.

Thus, co-ordinated and well-documented procedures are obviously of paramount importance if items are not to escape inspection. The coding system as used in other contexts to identify a piece of equipment can be adopted for the inspection control purposes and should be painted on to the item itself, to aid identification. In addition, a colour coding for each period of statutory inspection helps to ensure that equipment meets the stipulated periods. Notwithstanding the voluntary or insurance company inspection procedures installed by a machine user or owner, the Health and Safety Executive may inspect equipment at any time and suspend its usage if defects are present and/or certificates are not up to date. Maintenance and inspection records, can be effectively monitored and updated using standard or customised computer packages mounted on personal, networked or mini computers.

Stock control and spare parts policy

Stock control can play an important role in securing the effectiveness of a maintenance system. Manufacturers, suppliers and transport systems are rarely able to deliver goods at the exact time required for the maintenance operation, and it is therefore necessary to carry sufficient stock to act as a buffer between supply and demand for a component. However, since the level of stock is only a buffer, it is important to keep levels to the minimum needed to service the maintenance requirements and so limit the locked-up capital, which otherwise could be more usefully employed elsewhere in the business.

The extent to which component types and stock levels are held will often depend upon the nature of the maintenance policy and the proximity of the manufacturers' distributors. For example, the fleet operator hiring out to the market would probably carry a sophisticated range of spares, whereas the business with only a few items would hold only those parts in frequent demand. Thus, the extent to which stock control is made effective is dependent upon:

1 defining a realistic stock objective in relation to the firm's activities;
2 using stock as a buffer only to aid production continuity;
3 setting economic levels of stock to service the needs of the enterprise.

It is not possible to offer firm advice in setting the correct stock control objectives, without having detailed information for a particular concern. However, the following techniques are available for dealing with the problems arising in items (2) and (3) above: (a) the ABC method of stock control and (b) inventory control.

The ABC method

The ABC technique directs effort towards an ordering of stock priorities with the primary objective of avoiding stock-outs of critical items and keeping capital lock-up as low as possible. To operate the technique, stock is classified into three groupings:

- 'A' *items* are those which are most frequently demanded or expensive or which would have a significant impact if not provided;
- 'C' *items* are those not commonly requested, or which are inexpensive or would have the least effect if not provided;
- 'B' *items* are those which do not fall into categories 'A' or 'C'.

Clearly, for the stock control system to be effective, sophisticated information including forecasts of requirements, and lead times for ordering, will be required on 'A' items. 'B' items will require vigorous stock checking and frequent inspection, with more effort given to the high-priority items than to those of low priority. 'C' items need only routine checking, since the effect of a stock-out or miscalculation of the stock requirement will be relatively small.

Typically, stock items replenished over a common period could fall into the following categories:

- 'A' items: 10 per cent in number, making up 70 per cent of the value of stock;
- 'B' items: 20 per cent in number, making up 20 per cent of the value of stock;
- 'C' items: 70 per cent in number, making up 10 per cent of the value of stock.

Table 9.1 ABC stock control example

Class	*Number of items*	*Value of stock (£)*	*ABC value of stock (£)*
A	50	21,000	3,500
B	100	6,000	4,000
C	350	3,000	3,000
	500	30,000	10,500

ABC stock control example

The spare parts department of a plant hire company currently replenishes all stock items every three months, as detailed in Table 9.1.

The total cost of maintaining a uniform quantity of three months' minimum stock of all items is £30,000. However, with the items now clearly ranked into an order of stock priorities, decisions can be taken regarding more economic replenishing periods. For example, by keeping two weeks' stock of class 'A' items, two months' stock of class 'B' items and three months' stock of class 'C' items, the stock value reduces to £10,500. In this way capital is made available for other purposes.

It can be seen that the ABC method principally identifies the critical items, and so emphasis may be concentrated on checking class 'A' items to avoid stock-outs or supporting too much stock.

Inventory control

While the ABC technique of stock control provides a simple checking method, it is usually also necessary to know when to order and the order quantity. In its simplest form, stock is ordered when the current level of stock, minus the immediate stock demand, equals the forecast demand before the next delivery arrives plus safety stock, thus:

$$\text{Current stock} - \text{immediate stock demand} = \underbrace{\text{forecast demand}}_{\text{(order quantity)}} + \text{safety stock}$$

The safety stock for a given lead time may be determined from past observations of differences between forecast demand and actual demand. The size of the forecast error, which may only be exceeded within a specified limit, defines the size of each safety stock to provide against a stock-out. Unfortunately, this method requires information on each stock item, which for a complex inventory would be too tedious to collect and so only Class 'A' items might be considered in this way.

Method of determining the safety stock level

1 For each stock item calculate the forecast error by subtracting forecast demand from actual demand over a past time period. A positive difference represents a shortage, while a negative value indicates a surplus.

2 Rank the range of forecast errors into 10–20 equal divisions, for example:

Range of errors	Forecast error frequency
−24 to −20 (items, quantity, etc.)	5 (occurrences)
−19 to −15	20
−14 to −10	40
−9 to −5	60
−4 to 0	75
0 to +4	80
+5 to +9	65
+10 to +14	40
+15 to +19	15
+20 to +24	5
	405

3 A plot of the forecast errors against the frequency will tend to approximate to the normal distribution and simple statistics can be used to determine the probability of stock-outs and surpluses.

Example 9.1

The forecast errors shown above were recorded for a particular stock item. Management wishes to know with 95 per cent confidence the level of safety stock required to avoid a stock-out.

Solution

A plot of the forecast errors is shown in Figure 9.6. The sample is large and therefore approximates to a normal distribution, and the mean and standard deviation are calculated in Table 9.2.

$$\text{Mean stock error } (\mu) = \frac{\Sigma fx}{\Sigma f} = \frac{-40}{405} = -0.099$$

$$\text{Standard deviation } (\sigma) = \sqrt{\frac{\Sigma f(x-\bar{x})^2}{\Sigma f - 1}} = \sqrt{\frac{33{,}184}{404}} = 9.06$$

The data described by the normal distribution is governed by

$$z = \frac{x \pm \mu}{\sigma} = \frac{\text{safety stock}}{\sigma}$$

From statistical tables, for a given error value x to lie within 95 per cent of the area under the graph, $Z = 1.96$, that is, within 1.96 standard deviations of the mean. Therefore, safety stock $= 1.96 \times 9.06 = 17.76$ (say 18). If stock were ordered in four week cycles, the probability is that a stock-out will occur every 80 weeks, that is, 5 per cent chance.

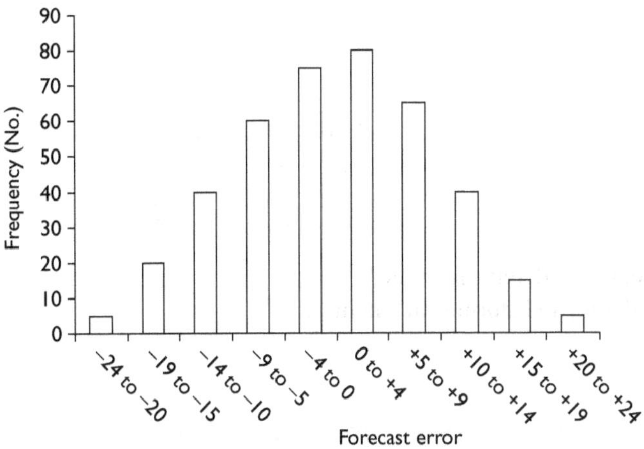

Figure 9.6 Distribution of forecast errors.

Table 9.2 Inventory control for Example 9.1

Range of errors	*Mid point of range, x*	$x-\bar{x}$	*Frequency (f)*	$f \times x$	$f \times (x-\bar{x})^2$
−24 to −20	−22	−22.1	5	−110	2,442
−19 to −15	−17	−17.1	20	−340	5,848
−14 to −10	−12	−12.1	40	−480	5,856
−9 to −5	−7	−7.1	60	−420	3,024
−4 to 0	−2	−2.1	75	−150	331
0 to +4	+2	+1.9	80	+160	289
+5 to +9	+7	+6.9	65	+455	3,094
+10 to +14	+12	+11.9	40	+480	5,664
+15 to +19	+17	+16.9	15	+255	4,284
+20 to +24	+22	+21.9	5	+110	2,352
			405	−40	33,184

Economic order quantity (how much to order?)

While the order quantity formula already given is appropriate for simple situations, for more complex stockholdings the costs of procurement and storage can be high. In such circumstances a more general formula is necessary relating order quantity, order period and costs. Two costs must be considered:

1 *Cost of procurement*: Administration costs are incurred whenever an order is placed. These will include (a) purchase enquiries, requisitions and ordering; (b) acceptance, inspection and legal requirements; (c) preparation of drawings and design details, etc.

2 *Cost of storage*: This includes (a) interest to be paid on working capital; (b) insurance; (c) storage and handling; (d) maintenance records; (c) wastage, theft, deterioration, etc.

The most economical ordering quantity involves balancing these costs against the rate of usage.

Example 9.2

Stocks of a component are allowed to run down to a level of three units before being replenished. The components are used steadily at 50 items per week. The component costs £15 per unit and the cost of storage and deterioration per week is 10 per cent of the cost price. Each time a component is ordered there is a cost of processing this order of £1.

1 How many items should be ordered each time?
2 What is the cost of ordering and storing each item?

Solution

Let B = minimum stock level, that is, safety stock. Let Q = number of components delivered with each order, D = rate of usage in units per week, S = cost of processing an order, h = cost of storing an item per week as a percentage of cost price, P = cost of an item and t = time in weeks between orders.

The cycle of usage and replenishment is shown in Figure 9.7. The average number of items stored in time t is $\frac{1}{2}Q + B$. Therefore, storing cost per cycle of length $t = \frac{1}{2}QthP + BthP$.

$$\text{Total cost per cycle of length } t = \tfrac{1}{2}QthP + BthP + S$$

$$\text{Total cost per week} = \tfrac{1}{2}QhP + BhP = \frac{S}{t}$$

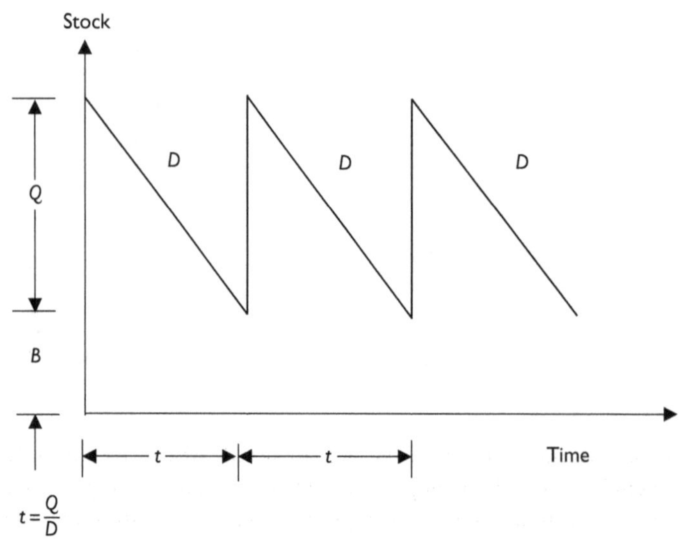

Figure 9.7 Cycle of usage and replenishment for Example 9.2.

But $t = Q/D$

$$\therefore \text{ Total cost (TC) per week} = \tfrac{1}{2} QhP + BhP + \frac{SD}{Q} \qquad (1)$$

To obtain optimum order size, differentiate with respect to Q:

$$\frac{\mathrm{dTC}}{\mathrm{d}Q} = \tfrac{1}{2} hP - \frac{SD}{Q^2}$$

$$= 0 \text{ for a maximum}$$

Therefore

$$Q^2 = \frac{2SD}{hP}$$

(1) $$Q = \sqrt{\frac{2SD}{hP}}$$

$$= \sqrt{\frac{[2 \times 1 \times 50]}{[(1/10) \times 15]}}$$

$$= 8.16 \text{ components} \qquad (2)$$

(2) Substituting Q in Equation (1)

$$TC = hPB + \frac{Q^2hP + 2SD}{2Q}$$

$$= hPB + \frac{2SD + 2SD}{2 \times \sqrt{2SD/hP}}$$

$$= hPB + \sqrt{2SDhP}$$

$$= (1/10) \times 15 \times 3 + \sqrt{[2 \times 1 \times 50 \times (1/10) \times 15]}$$

$$= £16.75 \text{ per week} \qquad (3)$$

Example 9.3: Stock control and shortages

A rental firm calls for the steady supply of 50 components each week; the price of the component is £30. The supplier usually keeps sufficient stock to meet demand, and the cost of holding a component per week is 10 per cent of cost price. The cost to the supplier each time a new order is processed is £10. However, sometimes the delivery date cannot be met, so to make up the backlog there are special deliveries to the customer as soon as the supplier is able to continue with the order. The extra cost incurred by the supplier in this situation is £10 per component.

1 calculate the economic order quantity for the supplier;
2 calculate the total cost per week to the supplier of stockholding and processing orders;
3 calculate the level to which stock on site is topped up.

Solution

Z = cost of storage per component, Q = economic order quantity, A = top-up quantity; D = rate of usage per week, h = storage cost as a percentage of the cost price of components, P = cost of components, S = cost of processing an order; t = time in weeks between suppliers (see Figure 9.8).

From Figure 9.8, storing as well as storage cost per cycle are calculated as shown below.

(1) Storing cost $= \frac{1}{2} Dt_1 t_1 hP$ per cycle and $A = Dt_1$

$$\text{Storing cost per week} = \frac{Dt_1^2 hP}{2t}$$

$$= \frac{At_1 hP}{2(t_1 + t_2)} \tag{1}$$

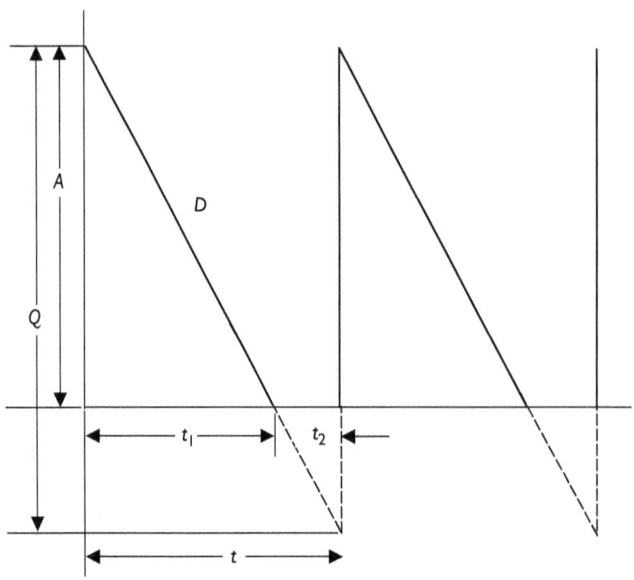

Figure 9.8 Stock control for Example 9.3.

Now, from similar triangles,

$$\frac{Q}{t_1 + t_2} = \frac{A}{t_1}$$

Therefore,

$$t_1 = \frac{At_2}{Q - A}$$

Substituting t_1 in Equation (1),

$$\text{Storing cost per week} = \frac{A^2hP}{2Q} \qquad (2)$$

(2) Cost of storage per cycle $= \frac{1}{2} Dt_2^2Z$

$$\text{Storage cost per week} = \frac{Dt_2^2Z}{2t}$$

$$= \frac{Dt_2^2Z}{2(t_1 + t_2)} \qquad (3)$$

Substituting t_1 in Equation (3),

$$\text{Shortage cost per week} = \frac{(Q - A)^2Z}{2Q} \qquad (4)$$

(3) Total cost per week $= \dfrac{A^2hP}{Q} + \dfrac{(Q - A)^2Z}{2Q} + \dfrac{SD}{Q}$

To obtain optimum order size, differentiate with respect to Q and A and maximise.

$$Q = \sqrt{\frac{2SD}{hP}} \cdot \sqrt{\frac{Z + hP}{Z}} \text{ units}$$

$$A = \sqrt{\frac{2SD}{hP}} \cdot \sqrt{\frac{Z}{Z + hP}} \text{ units}$$

$$TC = \sqrt{2SDhP} \cdot \sqrt{\frac{Z}{Z + hP}} \text{ per week}$$

1 $Q = \sqrt{\dfrac{2 \times 10 \times 50}{(1/10) \times 30}} \times \sqrt{\dfrac{10 + (1/10) \times 30}{10}} = 20.8$ components

2 $TC = \sqrt{[2 \times 10 \times 50 \times (1/10) \times 30]} \times \sqrt{\dfrac{10}{10 + (1/10) \times 30}} = £48.04$ per week

3 $A = \sqrt{\dfrac{2 \times 10 \times 50}{(1/10) \times 30}} \times \sqrt{\dfrac{10}{10 + (1/10) \times 30}} = 16.01$ components

Example 9.4: Stock control and discounts

Components are required at the rate of 2,000 per month. The cost of ordering is £20 and the cost of storing the material is 50 per cent of its purchase cost. The cost per item depends on the total quantity ordered as follows: (1) less than 500 items @ £1.21 per item; (2) 500–999 items @ £1.00 per item; (3) 1,000 or more @ £0.81 per item.

Calculate the optimum order quantity and the optimum total cost per month of purchasing, storing and ordering the material.

Solution

$$Q = \sqrt{\frac{2SD}{hP}}$$

$$= \sqrt{\frac{2 \times 20 \times 2{,}000}{\frac{1}{2}P}}$$

$$= \frac{400}{\sqrt{P}}$$

Without discounts, only the costs of storage have been involved in the calculation and not the cost of the material itself. This may result in a false situation, as follows:

$$(1) \quad Q = \frac{400}{\sqrt{1.21}}$$

$$= \frac{400}{1.1}$$

$$= 363 < 500 \text{ items (i.e. within range)}$$

$$(2) \quad Q = \frac{400}{\sqrt{1}}$$

$$= 400 \text{ (i.e. outside range } 500-999 \text{ items).}$$

$$(3) \quad Q = \frac{400}{\sqrt{0.81}}$$

$$= \frac{400}{0.9}$$

$$= 444 \text{ (i.e. outside range 1,000 items or more)}$$

On first inspection the optimum order quantity would be 363 units, but the calculation so far does not take into account the cost of the material itself, which in this instance varies according to the quantity ordered. Therefore, looking at total costs:

$$\text{TC} = \text{cost of material} + \text{storage} + \text{order cost}$$

Therefore,

$$\text{TC per month} = DP + \tfrac{1}{2}QhP + \frac{SD}{Q}$$

Overall discount situation:

(1) $\text{TC} = 2{,}000 \times 1.21 + \left(\tfrac{1}{2} \times 363 \times 0.5 \times 1.21\right) + \dfrac{20 \times 2{,}000}{363}$

$= \text{£}2{,}640$ per month

(2) (a) $\text{TC} = 2{,}000 \times 1.0 + \left(\tfrac{1}{2} \times 500 \times 0.5 \times 1.0\right) + \dfrac{20 \times 2{,}000}{500}$

$= \text{£}2{,}205$ per month

(b) $\text{TC} = 2{,}000 \times 1.0 + \left(\tfrac{1}{2} \times 999 \times 0.5 \times 1.0\right) + \dfrac{20 \times 2{,}000}{999}$

$= \text{£}2{,}289$ per month

(3) (a) $\text{TC} = 2{,}000 \times 0.81 + \left(\tfrac{1}{2} \times 1{,}000 \times 0.5 \times 0.81\right) + \dfrac{20 \times 2{,}000}{1{,}000}$

$= \text{£}1{,}862$ per month

(b) $\text{TC} = 2{,}000 \times 0.81 + \left(\tfrac{1}{2} \times 2{,}000 \times 0.5 \times 0.81\right) + \dfrac{20 \times 2{,}000}{2{,}000}$

$= \text{£}2{,}045$ per month

The optimum order quantity, therefore, is 1,000 items per month and the total monthly purchase, storage and ordering cost is £1,862.

Example 9.5: Stock control and continuous usage

A manufacturer is required to supply 1,000 parts each week to replenish stocks. The store has very little storage space and thus requires the units to be delivered at the rate at which they can be used. The manufacturer has the capacity to produce 2,500 units per week. The cost of storing a unit per week is 1p and the cost of setting up the equipment for a production run is £50.

1 What is the optimum number of units to produce in a production run?
2 What is the total cost of producing and storing the plant department's requirements?
3 How frequently should production runs be made?

Solution

(1) Q = number of units made per production run, D = number of units required by plant owner each week, k = number of units produced per week, H = cost of storing one item

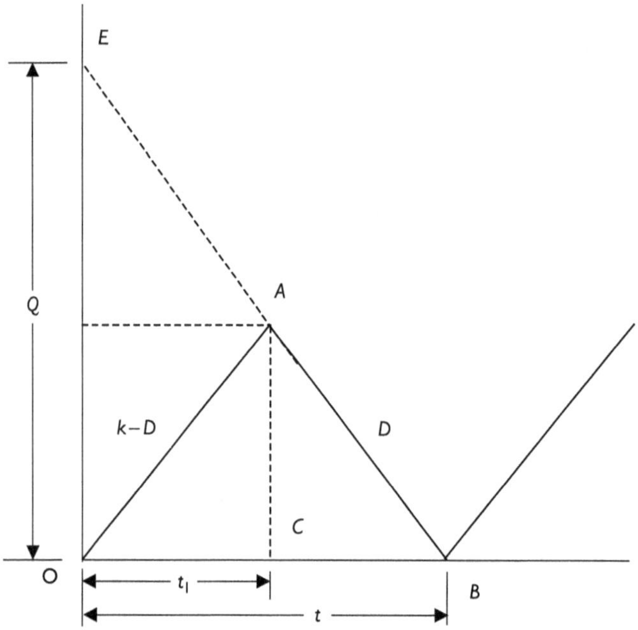

Figure 9.9 Stock control for Example 9.5.

per week, S = cost of setting up a production run, t = time interval in weeks between production runs (refer to Figure 9.9)

From Figure 9.9,

Length of the production run $t_1 = \dfrac{Q}{k}$

Length of production and usage cycle $t = \dfrac{Q}{D}$, $AC = Q - Dt_1$

Storage cost per cycle $= \frac{1}{2}(Q - Dt_1) \times H \times t = \frac{1}{2} Qt(1 - D/k)H$

Total cost per cycle $= \frac{1}{2} QtH(1 - D/k) + S$

Therefore

$$\text{Total cost (TC) per week} = \frac{S}{t} + \tfrac{1}{2} QH(1 - D/k)$$

$$\text{TC} = \frac{SD}{Q} + \tfrac{1}{2} QH(1 - D/k)$$

To obtain optimum run size, differentiate with respect to Q:

$$\frac{\text{dTC}}{\text{d}Q} = -\frac{SD}{Q^2} + \tfrac{1}{2} H(1 - D/k) = 0 \text{ or a maximum}$$

$$Q = \sqrt{\frac{2SD}{H[1 - (D/k)]}} \text{ units}$$

$$t = \sqrt{\frac{2S}{DH[1 - (D/k)]}} \text{ weeks}$$

$$\text{TC} = \frac{2SD + Q^2H[1 - D/k]}{2Q} \text{ per week}$$

Or, substituting for Q,

$$\text{TC} = \sqrt{[2SDH[1 - (D/k]]} \text{ per week}$$

$$Q = \sqrt{\frac{2 \times 50 \times 1{,}000}{0.01 \times [1 - (1{,}000/2{,}500)]}} = 4{,}083 \text{ units per run}$$

(2) $\text{TC} = \sqrt{(2 \times 50 \times 1{,}000 \times 0.01[1 - (1{,}000/2{,}500)])} = £24.50$ per week

(3) $t = \sqrt{\dfrac{2 \times 50}{1{,}000 \times 0.01[1 - (1{,}000/2{,}500)]}} = 4.08$ weeks

These examples clearly involve a refined level of stock control, necessitating detailed information gathering. For many equipment stores departments, the types of stocks and the spares are too few, and the usage too low, to need such sophistication. In that circumstance, adoption of the ABC technique coupled with rigorous checking of both stock and order processes should suffice.

Chapter 10

Health and safety

Introduction

The incorrect or inappropriate use of off-highway plant and equipment can result in serious injury, or in the extreme, fatality. Post Second World War, individual governments across continental Europe have recognised the issues of health and safety but until quite recently, had each developed their own legislation in an attempt to reduce risk of injury or ill health. Since the development of the European Union (EU), the Council of Ministers of the European Community (EC) has sought to harmonise this collection of disparate national legislation, through the introduction of EC Directives. The underlying rationale has been to equip a potentially transient workforce with consistent health and safety knowledge that ensures high safety standards are maintained as they migrate across EU member country boundaries.

More recently, EC Directives have been specifically targeted at mechanical plant and equipment usage. Such directives have been introduced and enforced. However, legislation *per se* cannot secure a safer working environment, but rather, it is the adoption of safe working practice that is important. Where the plant operator, maintenance crew and plant manager are not fully conversant with current health and safety requirements, then production processes become unduly dangerous.

The repercussions of accidents in the workplace are traumatic and often resonate throughout the organisation, through industry and society. Safety management is everyone's business, and no one engaged with plant and equipment usage should be excluded. Management should be conversant with the potential dangers faced by employees when operating with or near plant and equipment, in order that effective health and safety systems and procedures may be developed and implemented. Employees should assist employers in the implementation of these systems, procedures and good practice in order to ensure their effectiveness. Commitment of both the employer and employees to the organisational health and safety policy (together with effective communication of information) is the key to reducing accident rates. As the old adage goes – *know safety, no accidents*!

Given the wide range and diversity of legal requirements and supporting documentation, it would be difficult to address all health and safety legislation relating to equipment management in this one chapter. Therefore, this chapter introduces the principles upon which safe procedures and practice can be developed and applied within industry. Information provided should not be treated as definitive but rather taken as a general practical guide. It is strongly advised therefore that the original and relevant legislation is reviewed before developing plant and equipment safety systems and procedures.

The cost of health and safety

When developing (or updating) a health and safety policy relating to plant and equipment, management should give careful consideration to the cost implications involved. It is the moral duty and responsibility of all plant professionals (both employers and employees) to ensure the health, safety and welfare of themselves and others. This duty is backed and enforced by relevant legislation that has been developed to combat the growing incidence of plant related accidents. It therefore makes economic sense for a business to ensure that no unnecessary harm comes to any person as a result of plant usage, or from maintenance activities conducted on plant.

The costs of implementing a health and safety policy

Development and implementation of an effective health and safety policy will inevitably impose financial costs upon an organisation, whether directly through money spent on the strategy, or indirectly through time spent administering it.

Direct costs

The direct costs of implementation are more readily quantifiable and include resources to support the strategy, such as:

1 training and certification costs;
2 protective wear;
3 modifications to plant and machinery;
4 erection of site signs and the development of safe access routes;
5 insurance policies;
6 plant maintenance costs;
7 support services (administration, used oil analysis, workshops, etc.).

Indirect costs

The indirect costs associated with a health and safety policy are less tangible but no less significant. These include:

1 management time to plan, organise, deploy and administer the policy;
2 production of literature to communicate the developed policy;
3 employee time to learn, incorporate and support the strategy;
4 non-productive time spent on acquiring training.

The costs of not implementing a health and safety policy

Despite the seemingly high costs of establishing a health and safety policy within an organisation, the financial risks of not doing so may be far higher. Workplace injury can be expensive to the organisation and seriously damage business reputation and thus economic growth and prosperity. For example, in the US, construction accidents alone cost over $17 billion annually. Similarly, in the UK, a recent government estimate stated that the cost of health and safety failure (across all industries) could be as high as £18 billion every year (UK sterling, as at June 2000).

Direct costs

In the event of an accident occurring, the typical direct costs of not having implemented an appropriate health and safety strategy include:

1 reparative work to damaged plant and machinery or other facilities (buildings etc.);
2 criminal conviction resulting in fines;
3 compensation paid to the victims;
4 increased insurance premiums.

Indirect costs

Other indirect costs of not having an appropriate health and safety plan in the event of an accident include:

1 reduction in employee output as a result of illness or injury;
2 lost production due to inactive plant;
3 management time spent undertaking accident investigations and developing new prevention strategies;
4 disruption to work in progress leading to late completion and potential loss of future contracts;
5 bad corporate (and individual) publicity and possible imprisonment.

Given the previous, management must make a calculated assessment (usually via a cost benefit analysis) as to whether the costs associated with measures taken to ensure safe plant operation are reasonable and practicable given the risks involved. If not, then management would be strongly advised not to tender for the works. In the longer term, the cost of accidents far outweighs the cost of attempting to prevent them.

Accident statistics

There are various types of plant-related health and safety statistics available. However, as with any source of statistics there are several ways in which they can be interpreted (or misinterpreted). Table 10.1 provides general information on the total number of accidents/ incidents for earth moving machinery (e.g. dozers, dumpers, excavators, loaders and so forth) within the UK during the period 1986–1996.

The accidents are divided into five main categories: total accidents, fatalities, major injury, operative incapacitation for over three days and dangerous occurrence. From these statistics, it would appear that excavators are the most hazardous machine type to operate; with over 486 accidents in total and 21 fatalities over the ten year period. However, there are more excavators utilised throughout industry than any other machine type. This is because excavators are extremely adaptable machines that can be fitted with a diverse range of attachments including grabs, impact hammers and concrete crushers. Because of this versatility, excavators are the biggest selling machine type with around 5,000 sold within the UK alone and around 15,000 in the US. So although this table offers some insight into plant related accidents, it fails to indicate the underlying rate, that is, accidents as a proportion of the number of machines per category of machine. For example, 10 accidents for 100 site dumpers is proportionally higher than 80 accidents for 1,000 telehandlers.

Table 10.1 Total number of accidents/incidents per machine type in the UK, 1986–1996

Machine	*Number of accidents*				
	Total	*Fatal*	*Major*	*Over 3 days*	*Dangerous occurrence*
Dozer/bulldozer	35	6	14	8	7
Dumper	179	15	72	92	
Excavator	486	21	43	49	373
Grader	5		2	2	1
Landfill Compactor	6	1	1	4	
Loader	30	5	18	6	1
Pipelayer	4		2	1	1
Scraper	5	1	4		
Trencher	2				2

Source: Male (1998).

Table 10.2 illustrates the distribution of accidents by industrial activity over the same ten year period 1986–1996. This therefore highlights particularly dangerous industrial sectors. The construction industry appears to be the most dangerous although again, consideration should be given to the size of the industry and the number of plant items operating within it, that is, the 'average' needs to be observed. The Health and Safety Executive of the UK have recently confirmed the hazardous nature of the construction industry. Specifically, it revealed that the construction operative is seven times more likely to be involved in an accident than an operative in any other sector except mining. An operative in mining is twice more likely to incur an accident than an operative in construction.

In order to address the shortcomings of the statistics discussed so far, recent research has sought to provide a more in-depth appreciation of health and safety incidents within the construction industry This has been in order to determine not just the rate of plant related accidents, but also to observe whether the rate has decreased.

There are 20 categories of accident recorded in the statistics held by the Health and Safety Executive (HSE) (such as asphyxiation, exposure to fire, exposure to explosion, etc.), two of which could be directly attributed to plant and equipment usage. These categories are '*contact with moving machinery or material being machined*' and '*struck by a moving vehicle*'. The data available for accidents in these two categories is represented graphically in Figure 10.1. Here the trend in plant related accidents for all construction operatives between 1989 and 1999 can be seen; where the number of construction operatives has been divided by the accidents per annum to obtain an accident rate per 1,000 operatives.

Figure 10.1 shows a consistent trend throughout the period, with the average (mean) being 0.47 accidents per one-thousand construction operatives and the highest recorded rate occurring in 1989/90 (i.e. 0.55 per thousand). It is therefore apparent that the plant related accident trend for construction operatives has remained consistent and thus, largely unchanged throughout the period under scrutiny. This is despite the introduction of new European Health and Safety legislation. It would be erroneous to conclude that these findings provide definitive evidence as accident reporting (as opposed to accident occurrence)

Table 10.2 The distribution of accidents/incidents by industrial activity in the UK, 1986–1996

Machine	*Distribution of accidents (%)*								
	Quarrying	*Construction*	*Manufacturing*	*Other services*	*Transport*	*Agriculture*	*Education services*	*Disciplined services*	*Others*
Dozer/bulldozer	37	28	11	9	3	–	3	–	9
Dumper	6	83	4	3	1	1		1	1
Excavator	4	96	–	–	–	–	–	–	–
Grader	60	40	–	–	–	–	–	–	–
Landfill Compactor	33.33	66.66	–	–	–	–	–	–	–
Loader	23	77	–	–	–	–	–	–	–
Pipelayer	–	–	–	–	–	–	–	–	–

Source: Male (1998).

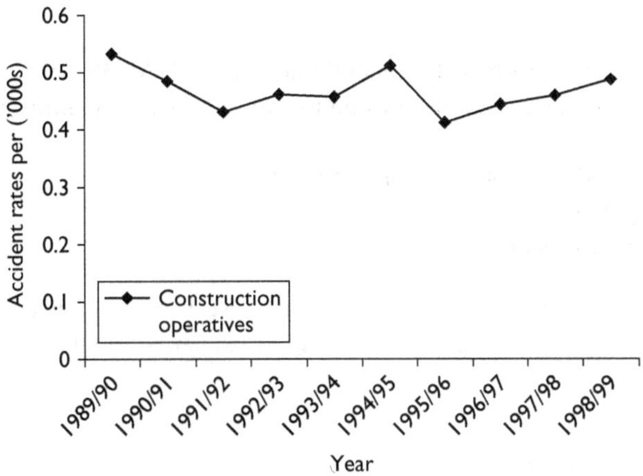

Figure 10.1 Accident rates per ('000s) construction operatives.

may have improved since the introduction of strict requirements stipulated by the Reporting of Industrial Diseases and Dangerous Occurrences Regulations 1995 (RIDDOR). However, despite RIDDOR, it should be recognised that although reporting has improved, many incidents are still not reported to relevant enforcement authorities. Nevertheless, the figures observed are very thought provoking.

Health and safety enforcement and guidance

Plant organisations, and the individuals operating within them, should not consider health and safety management as an extra business responsibility simply there to be complied with, but rather, as an integral and necessary part of daily work activities. There are mandatory requirements that must be adhered to but these should be perceived as an assistance to business activities, not a hindrance. After all, accidents not only maim and injure, but they also result in major losses in profitability as a result of downtime, maintenance costs and financial penalties.

To assist the practitioner to develop safe practices and procedures, a range of advisory bodies, legislative documents and practical notes are available. The eradication of all plant related accidents would be difficult to achieve. Nevertheless, through continual consultation with the relevant information sources, accident rates can certainly be reduced. The Health and Safety at Work Act (HASWA, 1974) provides a legislative framework designed to encourage high standards of health and safety at work. HASWA 1974 defines two authorities and gives them power for the enforcement of the legislation (Sections 10–14 and 18–24).

Health and Safety Commission

The Health and Safety Commission (HSC) was established under Section 10 of HASWA 1974. The Commission comprises one chairman and between six and nine lay commissioners representing industry, unions and local authorities. It is governed by the Secretary of

State and must adhere to directions provided; albeit the Commission does have a number of duties, these being:

1 to assist and encourage persons to further the general purposes of HASWA 1974;
2 to make arrangements for research and the promotion of training and information connected with it;
3 to disseminate information and act as an advisory service;
4 to submit proposals for regulations;
5 to approve and issue Codes of Practice.

The Commission has considerable powers to enable it to pursue these aforementioned duties effectively. These include directing investigations and inquiries, appointing staff and publishing information.

The Health and Safety Executive

The Health and Safety Executive (HSE) is a separate statutory body appointed by the HSC to work within its direction and guidance. The main duty of the Executive is to enforce the legal requirements and to provide an advisory service. It embraces an amalgamation of bodies, which had existed for some time prior to HASWA 1974 in a disparate manner, including the Factories Inspectorate, the Mines and Quarries Inspectorate and the Explosives Inspectorate. Breaches of HASWA amount to an indictable offence and the Executive has the power to prosecute offenders.

Health and Safety Inspectorate

The major legal requirements of HASWA 1974 are enforced through the work of HSE inspectors. These inspectors can enter premises without prior invitation and take samples, photographs and documents for the purposes of gathering evidence. Upon discovery of a contravention, they are empowered to take one of the following actions:

1 *Issue a prohibition or improvement notice.* A prohibition notice is issued if there is a serious risk to health or a risk of serious injury, and the activity giving rise to that risk must be subsequently stopped until remedial action is taken. Prohibition notices are served on persons undertaking the activity or on the person in control (i.e. business, premises, etc.).
2 *Issue an improvement notice.* An improvement notice is issued where a fault or contravention of a statutory requirement can be remedied within a specific time. The notice would be served on the person who is not satisfying the legal requirement or on any person on whom responsibility for health and safety has been placed, and this could be an employer, an employed person or even a supplier of equipment. There is a right to appeal to an independent tribunal against both a prohibition notice and an improvement notice.
3 *Prosecute the offender.* Prosecution can take place instead of, or in addition to, serving a notice. Failure to comply with a notice is also an offence against the law.
4 *Seize, render harmless and destroy a dangerous article.* Finally, the HSE inspector can seize, render harmless or destroy any substance or article that is the cause of imminent danger or serious personal injury.

Inspectors may also take written statements of fact from witnesses and ultimately conduct court proceedings even though they are not solicitors or counsel.

Legal documentation

There is voluminous health and safety documentation with which to consult, each with varying degrees of authority. Key documents can be classified as being either Acts, Regulations, Approved Codes of Practice or Codes of Practice. Each of these is now briefly discussed because an understanding of these documents is essential to health and safety management.

Acts

Acts (e.g. the Health and Safety at Work Act (HASWA 1974)), are known as statutes, which are passed by full parliamentary procedures and enforced by criminal law. Often, acts such as HASWA, called 'Enabling Acts', are arranged to allow supplementary regulations to be made by the Secretary of State without going through full parliamentary procedure. HASWA 1974 is wide ranging and covers everyone involved with work (both employers and employees) or those affected by it. In the USA, the Occupational Safety and Health Act (OSHA) affords similar protection.

Regulations

Regulations have the same power and status as Acts. To date, the majority of British safety regulations have been made under HASWA 1974. However, Britain is part of the EU which has sought to standardise legislation across member states. EC regulations now take precedence over national legislation; and recent years have seen the introduction of regulations such as the Provisional Use of Work Equipment Regulations (PUWER, 1998) and the Lifting Operations and Lifting Equipment Regulations (LOLER, 1998).

Approved codes of practice

Approved codes of practice (ACOPs) are documents which have been approved by the HSC, with the consent of the Secretary of State. Individual documents aim to provide practical advice to the practitioner regarding compliance with the law. That is, they define safe working practices and procedures (e.g. the Approved Code of Practice for the Safe Use of Work Equipment). ACOPs are not mandatory and there can be no prosecution for not following them. The law is flexible in that it allows alternative methods to those set out in the Codes to be utilised. ACOPs do however, have semi-legal status and thus in the event of an accident, failure to follow ACOPs may be viewed as a contributory factor in subsequent investigations.

Codes of Practice

Codes of Practice are guidance notes provided by trade unions and professional organisations and contain sound pragmatic advice and guidance. They often consist of a series of statements that prescribe the minimum standards of practice to be observed within a specific industry. Essentially, codes of practice are concerned with professional responsibility and are intended to be observed in the spirit and not merely the word. Similar to ACOPs, implementation of alternative 'bespoke' advice and guidance can be given as evidence in court.

Health and safety duties

The vast majority of health and safety legislation places a duty of care onto employers and employees to ensure that the health, safety and welfare of employees and the public are protected. Failure to comply with this duty of care is called negligence. For the lawyer there are three possible connotations for negligence. It may signify a person's state of mind, where they are blind to the consequences of their conduct. It can be a standard or measure of behaviour, for example, where a person has acted carelessly, either inadvertently or otherwise, and not as a prudent person would have done. Or finally, it may refer to the tort of negligence whereby a person has acted carelessly and exacted damage onto others and may be liable to pay compensation.

In order to protect themselves from negligent actions, plant owners and operators should understand the terminology regarding the required defined actions in the legislation. There are currently three generic duties:

1 *Shall* or *must* are absolute duties. These have to be obeyed regardless of cost and thus, if the duty is not feasible then the plant related activity must not take place.
2 *If practicable* means that the duty must be obeyed if feasible; cost is not a consideration. This definition of practicable is therefore more flexible than the previous duty. However, if management deem the duty not to be feasible and an accident results then proof of this assertion will be required.
3 *Reasonably practicable* requires a delicate balance between cost and risk and is more problematic. In the event of an accident the plant owner will be required to justify actions taken.

HASWA 1974 defines and imposes general duties upon a number of different categories of persons including employers, the self employed, those people who control premises, plant manufacturers and employees. This philosophy of shared responsibility is an important doctrine of health and safety legislation since it would be unrealistic to expect the individual to bear the brunt of all legislative requirements, or to solely ensure a safe working environment.

Employers

It is the duty of every employer to ensure, so far as is reasonably practicable, the health, safety and welfare at work for their employees. This applies particularly to the provision of equipment and systems of works, that are safe and without risk to health. In particular, the duties of the employer include:

1 ensuring that all machinery, equipment and appliances used are safe;
2 ensuring that the handling, transportation and storage of materials are safe;
3 providing information, instruction, training and supervision to ensure the health and safety of employees at work;
4 maintaining safe means of access to and egress from places of work;
5 provision and maintenance of a safe working environment with adequate facilities for welfare at work;
6 preparation and appropriate revision of a written statement explaining the employer's general policy with respect to the health and safety of employees, and organisation and arrangements made to carry out the policy effectively.

Employers also have a duty with regard to the health and safety of persons who are not employees, such as self-employed or contractor's employees who may be working with their own employees. This duty extends to members of the public and employers must ensure that the public are not exposed to health and safety risks.

Self-employed

The self-employed must ensure that as far as it is reasonably practicable, their conduct does not expose themselves, or others, to any risk from health and safety. This point is of particular relevance when considering the nature of plant hire, where machines can be hired 'driverless' to self-employed private contractors. Plant managers should note that the PUWER regulations state that the operator must be competent, but they do not stipulate that training is mandatory. In some instances, the self-employed have operated machines on site even though very limited training has been received. Practitioners have recognised this as a loophole in the legislation and have sought to enforce mandatory certification throughout industry via self-governed safety charters. The Major Contractors Group within the UK construction industry represents a typical example of industry self-government.

Those who control premises

Those who control premises (e.g. quarries and mines, construction and civil engineering sites and so forth) also have duties under HASWA 1974. They must ensure that as far as it is reasonably practicable, all means of access and egress are safe and that any plant or substance on the premises is without risk to health, with regards to those persons who use the premises as a work place.

Plant manufacturers

Plant manufacturers must ensure that articles and substances are so designed and constructed so as to be safe and without risk to health. The caveat here goes on to state 'when properly used', as the legislation cannot totally prevent the misuse of equipment and plant by incompetent plant operators. Neither can they prevent the influx of 'grey', non-CE marked machines that enter the EU each year. These are machines which are imported from countries outside the EU and do not conform to standards set under European law (e.g. with respect to emissions, noise pollution and so forth).

Employees

Employees must take reasonable care for their own and other persons' health and safety. For example, they should use safety equipment provided without interference, such as tampering with locks on movable guards. The employee must also co-operate with employers and any other relevant person(s) to enable them to perform their statutory duties. In order for them to achieve these objectives, employees must consult with their worker safety representatives and trade unions where applicable. If an organisation has five or more employees it must have a written safety policy defining its responsibilities and employees must be aware of its existence and content.

Health and safety management

The PUWER 1998 provide guidelines which aim to engender good practice. They state that account should be taken of:

1 working conditions and hazards encountered in the work place;
2 additional hazards created by the use or installation of work equipment;
3 equipment to ensure that it is fit for the intended purpose;
4 equipment maintenance to ensure that machinery is safe to operate;
5 operative (and other plant professional) training and instruction in order to ensure competence.

These directions provide a commonsense approach to plant and equipment safety management and are broadly applicable throughout most industrialised nations (not just those within the EU). The directions cannot prescribe how an organisation should develop its health and safety strategy since this would be impossible given the diverse range of industrial activities.

This section therefore provides a generic framework for a comprehensive health and safety management programme for plant and equipment. The guidance provided can be used by both plant professionals charged with the responsibility of developing a new health and safety programme and those evaluating and refining existing programmes. Information contained herein will serve, along with other supplementary reference materials, as an invaluable resource to assist plant professionals reduce the number and severity of plant related accidents in the workplace.

Management of a health and safety policy is a cyclical and continuous process that can be broken down into five core components (refer to Figure 10.2). These components are:

1 *Developing* (or redeveloping) an effective health and safety policy that provides clear direction for the organisation and its employees.
2 *Organising* the enterprise into an appropriate and efficient structure to facilitate the effective delivery of the health and safety policy.
3 *Planning* a rigorous and systematic programme of action to ensure the effective implementation of the health and safety strategy proposed.
4 *Evaluating* performance of the health and safety policy developed through comparative measurement against mandatory and in-house standards and procedures.
5 *Reviewing* performance through either an internal or external review or a combination of both.

Each core component relies upon the success of the others to ensure that the whole framework of health and safety management is successful. Greater consideration is now given to each stage.

Developing the health and safety policy

The health and safety policy should be written to complement the plant organisation's mission and goals and should reflect the values and beliefs of those who devised and implemented them. Management rhetoric should be avoided but rather, a genuine attempt to

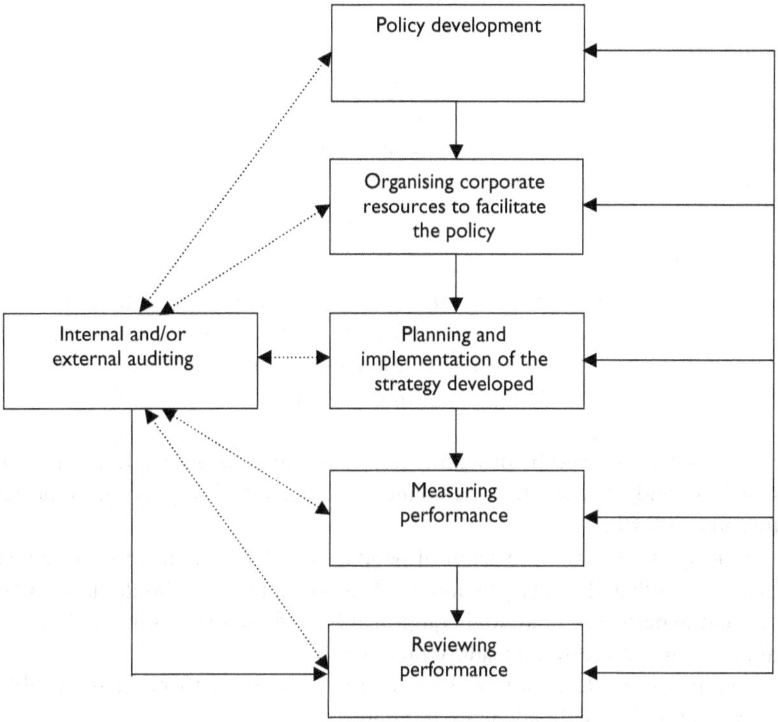

Figure 10.2 Core components of the health and safety management loop.
Source: HSE (2000).

engender and foster safer working practices throughout the organisation should be made. A typical policy would include:

1 *A statement of policy* that details who is responsible for health and safety management and establishes key performance indicators.
2 *A general statement of intent* that details what and who will be covered/included in the policy.
3 *Arrangements for the operation of this health and safety policy* which details how measures and procedures will be employed to reduce accident/injury occurrence.

It is important that employees, or their representatives, are consulted when developing the policy, because they are after all, an integral part of the process. This will help engender a sense of policy ownership throughout the organisation. Employees (in particular, plant operators) possess inherent knowledge of the operational environment and equipment they operate. This is particularly relevant in the construction industry because site managers often 'evolve' from either a building surveying, quantity surveying or construction management background and may therefore lack 'formal' plant and equipment knowledge. Consultation with the plant operator will prove to be an invaluable input source into plant aspects of a health and safety policy.

Organising business resources

Business resources should be organised so as to provide a framework of employee and employer commitment to the health and safety policy. The framework should support the policy and help with its implementation. Support mechanisms include management commitment, employee participation, appropriate systems of accident reporting, safety expertise and staff/employer competence. Organisation is therefore a multi-disciplinary, team-based exercise.

Management commitment

Management have control over most corporate resources, so their organisation of such must provide demonstrable evidence of unequivocal commitment to health and safety. This is achieved by compliance with safety legislation and by providing visible procedures, assistance and support to the efforts of others. In particular, managers should:

1 Endorse the policy developed by providing necessary implementation tools and support services such as budget, expertise, personnel with accountability, programme review procedures and so forth.
2 Empower a designated person, or team of people, with the responsibility to manage a comprehensive health and safety programme. Chosen candidates should have sufficient experience and expertise in plant and equipment hazard identification and have access to relevant and current health and safety literature.
3 Provide funding to enable health and safety representatives to purchase health and safety equipment and provide adequate training.
4 Ensure that professional development interviews/evaluations contain specific reference to health and safety management.
5 Create, communicate and enforce a control and disciplinary system that applies equally to all staff (plant operators, plant managers, site supervisors, etc.).

Employee participation

Employee participation and commitment to the policy should provide an effective means through which they can identify hazards, recommend and monitor abatement, and otherwise participate in their own protection. Participation in the decision-making process is of particular significance since it will help to empower and motivate employees to actively pursue policy objectives and goals. In order to facilitate employee involvement in the health and safety policy and moreover, organise this valuable resource to its best capability, management should:

1 Designate employees for health and safety tasks and duties that exploit their special interest and/or expertise.
2 Allow employees and their representatives to become involved in the inspection of plant and the operational environment. Furthermore they should be permitted to observe safety monitoring and obtain results in order to digest relevant material.
3 Develop procedures that actively encourage employees to promptly and accurately report complaints of hazards, unsafe work practices and occupational injuries and illness.

4 Ensure that employee participation is documented, for example, machine inspection reports, minutes from committee meetings and employee evaluation reviews.
5 Act promptly upon health and safety suggestions or concerns expressed by employees.

Accident, near miss and injury reporting

An effective method by which to prevent future accidents is to examine previous incidents of illness and injury, that is, to learn from examples of accidents and near misses in order to recognise their cause and thus prevent their reoccurrence. The organisation can learn to recognise common hazards (together with their possible outcomes) and more importantly, how to prevent them. It is therefore essential that adequate resources are made available for recording and analysing the accidents or 'near-misses' that do occur. An accident reporting procedure must therefore be rigorously followed for every injury or near miss, no matter how trivial, and recorded in the organisation's accident book.

The health and safety committee and safety representatives

Resources should be made available for safety representatives to implement good practice to their best ability. The elected safety representative is entitled to:

1 investigate potential hazards and dangerous occurrences at the workplace;
2 examine the causes of accidents and investigate employee complaints concerning health, safety and welfare at work;
3 make representations on any of such matters to the employer;
4 conduct health and safety inspections;
5 receive information from health and safety inspectors;
6 attend meetings of safety committees.

There is also permitted time off with pay, to perform the above tasks and receive adequate training.

Competence

Competence is a vital element in achieving success in the implementation and management of any health and safety policy. Essentially, competence is achieved by a combination of initial training, practical experience and accrued knowledge. This aspect of competence is discussed in greater detail in Chapter 11.

Planning and implementing the strategy

It is virtually impossible to design a fail-safe health and safety system. However, it is possible to identify potential hazards as a precursor to developing better standards and procedures that will invariably reduce accident occurrence. The procedure is known as risk assessment and represents a careful examination of potential sources of injury within the work environment so that companies (and individuals) can ascertain whether sufficient precautions have been taken or whether further (more exhaustive) measures are required.

Risk assessment is a legal requirement under most modern legislation (e.g. Management of Health and Safety at Work Regulations 1999). The assessment must consider a balance between the cost and complexity of the safety system, against the likelihood and severity of injury. For example, brake failures on articulated dump-trucks can kill but the likelihood is remote if regular maintenance work is conducted. Benefits of the risk assessment process are that it:

1 Assists inspectors, in the event of an accident, in determining whether reasonable, sufficient and appropriate measures have been taken to reduce the risk.
2 Raises awareness to both management and other stakeholders, with respect to particular hazards and/or precautions that may reoccur on a cyclical basis.
3 Helps determine whether legal requirements have been fulfilled.

As a general rule, the HSE advise that risk assessments should not be over-complicated but rather, that individuals should embrace a 'commonsense' approach. There is no need to be a health and safety expert; although on occasion, expert consultation can be invaluable. Risk assessment is covered in greater detail later in this chapter.

Evaluating health and safety performance

It is important to conduct regular measurement of the health and safety performance of the organisation and then evaluate it against key performance indicators to determine whether the strategy developed has been successful. That is, to determine whether accidents or reoccurrence of injuries, illness and damage to property have been prevented. Indicators usually include employee and supervisor training, finance to support the strategy, quality of record keeping, accident trends and so on.

There are two core elements to any credible monitoring system, namely passive and active surveillance. Passive surveillance enables an organisation to learn from existing statistics, either held in-house or obtained via external sources. For example, the HSE's Operations Unit in Bootle Merseyside provides an invaluable source of accident statistics and reports throughout industry. Active surveillance utilises a more proactive 'root cause' approach through the meticulous collection of data that is not currently documented. Questionnaires and workplace surveys are the prime mediums through which to obtain new information. Questions are usually structured so as to ask employees and management to identify potential hazardous practices and procedures and in addition, recommend measures to reduce these risks.

Reviewing and auditing performance

An organisation's health and safety policy arrangement must be fluid, that is, it must be able to evolve with experience. However, information is only as good as the application to which it is put; to record information and ignore it thereafter would be erroneous. Health and safety risk assessments must be regularly reviewed via an internal or external audit in order to assess the effectiveness or otherwise of actions implemented. That is, are the precautions working effectively? Moreover, the rate of mechanical technological development presently taking place means that new machines and operational and maintenance procedures will be introduced, which themselves will bring with them new hazards. It is not necessary to introduce new assessments for each and every trivial change in working practices or for every machine purchased, but nevertheless, some consideration must be given before any decision is taken. The cyclical review allows the organisation to reflect upon health and safety strategy and consequently, reinforce, maintain and further refine it in order to continually reduce risk.

Risk assessment

There are five steps to conducting a health and safety risk assessment, these are (refer to Figure 10.3):

1 look for hazards;
2 decide who might be harmed and how;
3 evaluate the risks and decide whether existing precautions are adequate or whether more should be done;
4 record the findings and precautions taken;
5 review the assessment and revise it if necessary.

Each step represents part of an iterative process that enables an organisation to acquire detailed knowledge of potential health and safety risks in the workplace.

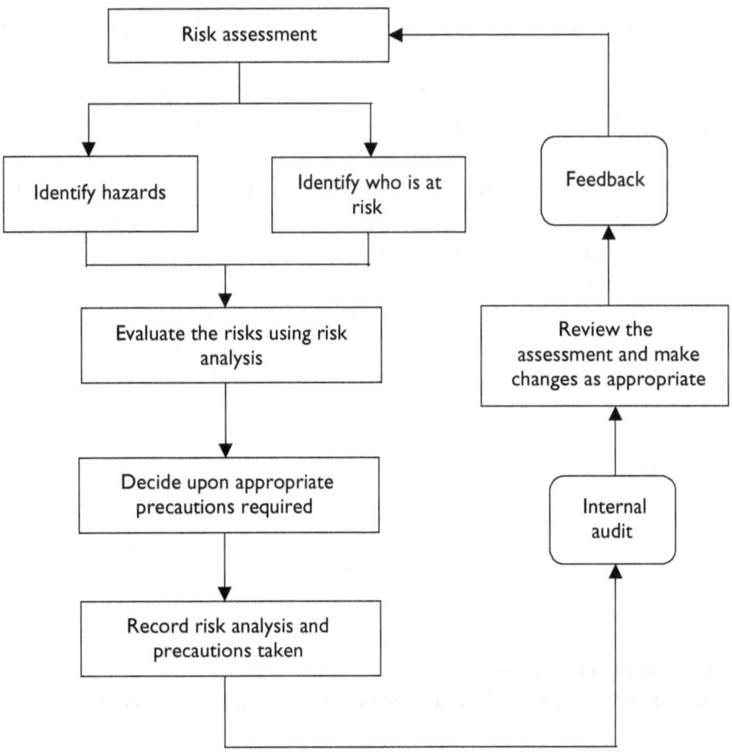

Figure 10.3 The risk assessment process.

Identify hazards

Hazard identification on machines or processes can be achieved by inspections, internal and external audits, study of incidents (near misses) and, for new plant, by investigation at the design stage, although this is largely the remit of plant manufacturers. Plant owners can reduce potential hazard by refusing to purchase substandard 'grey machines'. Involvement of the workforce at an early stage of the process (i.e. operators, maintenance crews and so forth) will help ensure that they adopt these safety procedures developed. It will also provide them with the opportunity to identify potential hazards that may not be immediately 'obvious' to management. Examples of hazards that can be reasonably anticipated include: crushing, temperature hazards (burning or freezing), impacting, snagging points which lead to entanglement, blinding, explosions via demolition/quarrying activities, cutting from moving edges, stabbing, shearing (leading to amputation), electrical hazards, inhalation or contact with dangerous materials and so on. Failure modes should also be considered, using standard methods such as a hazard and operability study.

Hazard and operability study

The hazard and operability study (HAZOPS) was created in response to a need to cope with the increasing number of plant items and industrial processes used within industry. Traditional methods of safety analysis based on compliance with standards, inspections and checklists cannot fully account for the complexity of modern working practices. HAZOPS represents a method by which to identify and document hazard and operability problems that could arise during plant and equipment operation in both normal and/or abnormal operation/ maintenance.

The HAZOPS exercise consists of two steps. The first step reviews existing health and safety systems together with the work environment. The second step consists of a follow-up study. The review team, comprising of operators and managers (and often chaired by a safety representative), is given a site layout together with a list of the plant and equipment utilised. A team member who is familiar with the equipment describes the application of specific plant and equipment items on site. The chairperson then leads the team through a list of discussion points to stimulate a debate on the possible causes and consequences for each deviation from the safety system that could occur. Deviations include rising temperature, failure to acknowledge safety on-board features such as alarms and so on. The causes, consequences, safeguards and any recommendations are recorded and informed actions taken.

Identify who is at risk

Consideration should be given to inexperienced workers, visitors/members of the public and other people who share the work place (tradesmen, site management, visitors, etc.) as well as experienced veteran workers. Common terms that should be considered, with specific definitions, are:

1. hazard (the potential to cause harm);
2. risk (a function of the likelihood of the hazard occurring and the severity);
3. danger (the risk of injury).

These terms enable management to quantify accidents that may occur.

Evaluate the risks using risk analysis

The third stage assesses each risk in turn as part of the physical risk assessment, which will document the severity of risk. There is no definitive method of conducting such evaluation as each plant item requires different levels of operator competence and maintenance work. Nevertheless, as generic guidance the inspector should consider: (1) the likelihood that the hazard could cause harm; (2) the possible severity of the accident and (3) the length of exposure to the hazard. It is important to note that in the event of an incident, the authorities will inspect the risk assessment conducted together with any other supplementary 'supporting' information.

(1) *The likelihood that the hazard could cause harm.* Considering the likelihood that a hazard could cause harm is an obvious but important aspect of the risk assessment, since this will determine whether or not the risk should be reduced. Even after all reasonable precautions have been taken some risk will usually remain. It is important to then decide whether this remaining risk is high, medium or low. Legislation generally provides the 'minimum requirements' and if at all possible, plant professionals should aim to minimise the risk by adding precautions as necessary. Ideally, hazards should be removed altogether if at all possible and if not, then adequate control mechanisms should be provided.

(2) *The possible severity of the accident.* When assessing the severity of the possible injury, various sources suggest the following four classifications:

(a) fatality, to include one or more deaths;
(b) major non-reversible injury, for example, amputation, loss of sight, disability;
(c) serious reversible injury but medical attention required, for example, burn, broken joint;
(d) minor small cuts and bruises.

Recording this data will enable the organisation to benchmark its health and safety performance against that of previous years as well as examine the magnitude of accidents that occur.

(3) *The length of exposure to the hazard.* The next step is to consider how often people are exposed to the risk. Suggestions here are:

- frequent – several times per day or shift;
- occasional – once per day or shift;
- seldom – less than once per week.

Also important is how long the exposure period will be. Is the person exposed to danger for a few seconds per event, or as can occur with major maintenance work, several hours? There may also be a need to consider the number of people who may be at risk; often a factor in petro-chemical plants.

Record risk analysis and precautions taken

From studying machine operation, the probability of 'potential' injury can be assessed as being either very certain, likely, probable, possible, unlikely or very unlikely. The method of assessment however, can vary. A popular option is to apply a points scoring scale to each of the factors above then use the total score to determine high, medium and low risks. In the following example, severity of the accident (S) uses a 50 point scale ranging from 5 for a minor scratch to 50 for a multi-fatality, the frequency of the hazardous activity (F) uses a scale of 1–10 and the probability of injury (P) uses a 5–50 scale. These three factors are combined to calculate the risk rating score (RRS) in the formula:

$$\text{RRS} = F \text{ (exposure to the risk `}x\text{')} \times [(S+P)\times 0.1] \text{ (possible harm resulting `}y\text{')}$$

The factors are then multiplied to determine a score between 1 and 100; where 1 represents wholly safe conditions and 100 represents totally hazardous conditions. Consider the following three scenarios:

Point (A) $F = 10$; $S = 50$; and $P = 50$. Therefore, $10 \times [(50 + 50) \times 0.1] = 10\ (x) \times 10\ (y)$ = risk rating is high at 100. Work must stop immediately.
Point (B) $F = 1$; $S = 5$; and $P = 5$. Therefore, $1 \times [(5 + 5) \times 0.1] = 1\ (x) \times 1\ (y)$ = risk rating is low at 1. Work can progress.
Point (C) $F = 5$; $S = 20$; and $P = 30$. Therefore, $5 \times [(20 + 30) \times 0.1] = 5\ (x) \times 5\ (y)$ = risk rating is medium at 25. Work can progress once further safety measures have been implemented.

In the case of possible harm resulting, decimal places are rounded up to the next whole number co-ordinate.

As an alternative, the factors of the *exposure to the risk* and the *possible harm resulting* (where the latter is obtained from the summation of the severity and the probability) provide x and y co-ordinates (on a 1–10 scale) which can be mapped against a prepared tabular format. This can be classified into high, medium and low risk (refer to Table 10.3).

Table 10.3 Risk rating: a tabulated comparison

Exposure to the risk (x)

Possible harm resulting (y)	1	2	3	4	5	6	7	8	9	10
1	B									
2										
3										
4					▲					
5					C			▲		
6	▲									
7										
8										
9										
10										A

Risk is low and no further action required.

Risk is medium and further improvements should be implemented.

Risk is high and no further work should be undertaken until the risk is reduced.

Both systems presented here are purely illustrative. The definition and delineation of the boundaries between the three classifications of high, medium and low risk exposure should be resolved by practitioners working in a given industry and when applying 'reasonable' judgement. The examples and accompanying table must therefore be adapted to given work situations and would not be appropriate for every work environment.

Review the assessment: the internal audit

The next stage of risk assessment is to devise, plan and implement methods, systems and procedures to reduce risks identified to an acceptable level. These may include reducing the risk by good design (e.g. the development of appropriate transportation routes both on and off site), adequate training and certification of plant operators, containment of rotating or reciprocating parts by guarding, provision of safety equipment and establishing written safe working procedures which must be followed. These should be backed by appropriate medical treatment, emergency procedures and fire precautions.

Recording the findings

If the employer has fewer than five employees, there are no mandatory requirements to record the original risk analysis or store data concerning actions taken to minimise the risk. However, it is useful to maintain a record of actions taken for future comparison and as evidence that in the event of an accident every reasonable precaution has been taken. For employers who employ over five employees, a written record of the significant findings of the assessment, together with hazards identified and conclusions made must be kept. Typically, this could include: type of, level attained and duration of training received by the operator; daily and workshop maintenance conducted; vehicle safety inspections, etc.

Health and safety legislation

There is a vast amount of legislation covering health and safety and to cover every item of legislation, applicable to each individual industry, within one textbook would be difficult. However, a list is given below of those which are commonly encountered within industry. This list is by no means definitive but readers are requested to seek out and read relevant legislation and appropriate Codes of Practice a priori to the development of individual plant and equipment health and safety strategies. Key legislation includes:

Management of Health and Safety at Work Regulations 1999;
Manual Handling Operations Regulations 1992;
Provision and Use of Work Equipment Regulations (PUWER) 1998;
Personal Protective Equipment at Work Regulations 1992;
Workplace (Health, Safety and Welfare) Regulations 1992;
The Control of Major Accident Hazards (COMAH) Regulations 1999;
Agriculture (Tractor Cabs) Regulations 1974;
The Lifting Operations And Lifting Equipment Regulations (LOLER) 1998;
The Construction (Design and Management) (CDM) Regulations 1994;
The Quarries Regulations 1999;
The Railways (Safety Case) Regulations 2000;
The Work in Compressed Air Regulations 1996.

Chapter 11

Investing in off-highway plant operator development

Introduction

The widespread deployment of plant and equipment in industry has unquestionably streamlined organisation efficiency and increased profitability. In an attempt to maximise plant productivity, management tend to focus most attention upon the machine and its capabilities, whilst less consideration is given to the impact of the operator utilising the machine. Yet, the interface between the human resource and the plant item is an essential aspect of plant management. Plant ownership *per se* does not translate into efficient production – the machine is only as good as the person operating it!

Operators share a symbiotic relationship with their equipment, employing a unique combination of physical and mental attributes during plant operation. Daily maintenance conducted on a machine by the operator is also important in order to maintain high machine utilisation levels and help ensure safe operation. Furthermore, the human resource is the only factor of production that competitors cannot replicate. Other resources such as materials and machinery are available to specified standards and can be 'purchased' given available finance and time. Therefore, operators and their ability are the most important differentiators in a competitive market.

To realise the maximum potential from any operator, an effective training and development plan is required. The plan should include a range of complementary training methods; including, theoretical training, practical training, on-the-job training and periodic re-evaluation. In combination, these methods are concerned with developing the operational capabilities of the operative to suit individual industrial applications. Neither should training be perceived as a one-off event, but rather, a part of a continual process. Experience accrued during this ongoing process will enable the operator to acquire competency.

It is important to note that the training of plant operation should not be restricted to operators alone. As part of continual professional development and the monitoring of operators themselves, the training instructors, plant foremen and plant managers should similarly receive a sound working knowledge of plant operation if a safe, productive and competitive culture is to be instilled within the workforce. After all, combined, these are the people who will monitor, control and enhance an operator's performance in practice in order to maximise profitability.

In this chapter, the fundamental aspects of operator development will be introduced and discussed.

Cost benefit analysis of training

Investment into operator training can initially be an expensive and time-consuming process and may not always appear to produce the 'expected' results (in terms of desired operator performance and ability). This perception derives from the fact that training costs occur within the short-term and whilst being trained, operators are not immediately productive; a further cost to the employer. In addition, the low rates of operator pay received and the cyclical nature of the industrial economy has partly created a transient plant operative workforce who migrate between industries in order to secure a continuation of, or increased, financial reward. Hence, it is important to retain trained operators because if an operator leaves an organisation once trained, the investment in that training is lost.

Most plant managers are aware of these implications and therefore give great consideration before investing finance into operator training. Indeed, a significant proportion of operators are still trained informally by their peers (i.e. by other more experienced operatives who are often not formally trained themselves). In such cases, operator knowledge and skill is largely acquired through trial, error and experience. Whilst this approach may seem practical and economical it fails to present a viable alternative to formal training, since there is little consensus, and much debate, as to what operator attributes actually determine competence.

Formal training represents an investment for the future, without which an organisation may not be able to meet the long-term needs of customers or changes due to new technologies. Once trained and competent, the employer should continue to nurture the operator's ability and skills and recognise the inherent professionalism of the trade. Rewarding the operator with a good wage, bonus incentive and working conditions will help ensure optimum performance and commitment to the organisation.

Pertinent, sensible operator development, backed by management support and encouragement, will enable operators to continually enhance their skills and knowledge. There are several important benefits to be accrued from improvement in an operator's performance (refer to Figure 11.1). For example, there will be;

1. *A reduction in accident occurrence* (measured both in terms of self-injury and injury involving other operatives): Adequate and appropriate training is undoubtedly the most critical element in any health and safety strategy and can help protect the plant operator's safety as well as the organisation's long-term survival.
2. *An increase in machine utilisation levels*: As a result of operators maintaining and operating their equipment to predefined standards set by a combination of the manager, legislative requirements and the plant manufacturer.
3. *A reduction in maintenance and running cost expenditure*: As a result of more efficient operation.

The culmination of these benefits will manifest themselves as increased productivity and higher profit margins for the employer. Such efficiencies could be passed on to the client in terms of lower production costs, timely completion of works or delivery of materials without the need to impinge upon quality.

Conversely, the negative ramifications of the poorly trained operator include reduced production output, greater potential risk to safety (both to operators and their site colleagues) and severe financial damages in the event of injury or damage to person or property. Arguably therefore, the efficient and effective management of plant operators is perhaps one of the most important areas of off-highway plant management. It is also one of the most neglected topics in most textbooks on plant and equipment management.

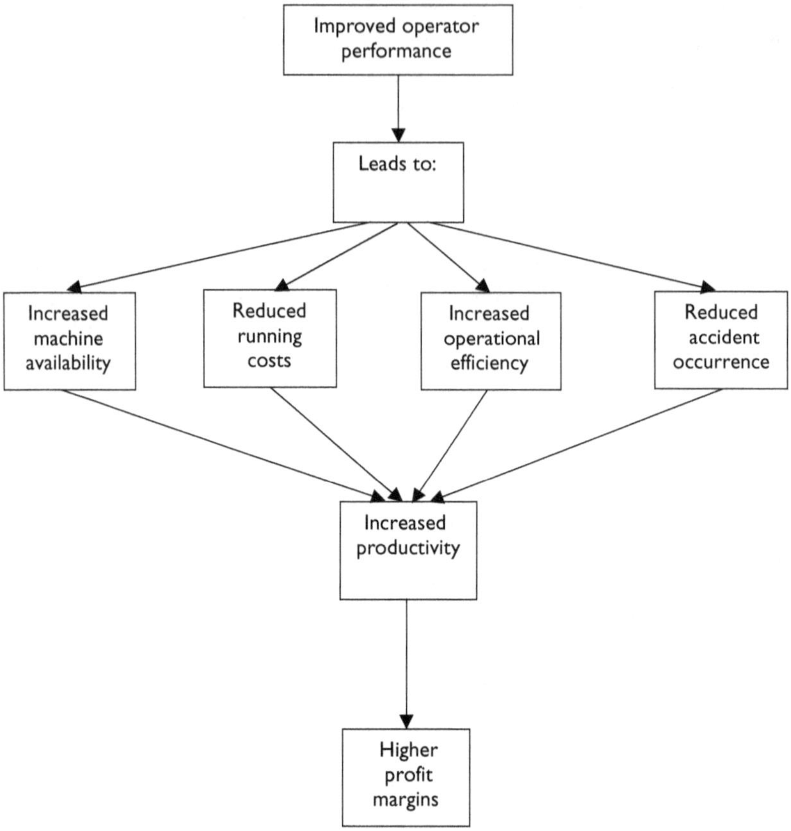

Figure 11.1 Benefits to be accrued from improvements to operator performance.

Health and safety implications

From the viewpoint of the client and employer, acquiring some protection from litigation resulting from a health and safety breach is essential. This is because in the event of an accident, legal proceedings may be taken against both client and contractor. If found guilty, the accused could receive either a substantial fine or imprisonment. As a result, many companies have opted to adopt rigorous systems and procedures (the infamously known *tick box culture*) in addition to formal training regimes employed in an attempt to protect themselves from any health and safety breach. Such a culture cannot indemnify practitioners from litigation, although it may serve to illustrate where further operator training is required.

Operator certification

Throughout the UK and European Union, training certification is not a mandatory requirement. Rather, the Provision and Use of Work Equipment Regulations (PUWER, 1998) state that an operator 'must be competent', but do not specify a means by which to obtain this competency. Such an 'open' approach has spurned the development of numerous bespoke training schemes and initiatives. Some purely train the operator to a core standard of knowledge (this being set by the individual training organisation), whilst others conduct training and issue some form of competency certification. Undoubtedly, most training organisations provide a suitably high standard of training material, delivered at an appropriate level, and costed to within affordable budgets. However, distinguishing the merits and demerits of individual training organisations and courses can be confusing and so there is a growing trend for organisations to supplement external training programmes or to train and educate totally in-house.

The Certificate of Training Achievement (CTA) card scheme (administered by the Construction Industry Training Board (CITB)) is the largest scheme available within the UK and has over 153,000 certificate holders. The CTA card currently issues competency based cards for road and rail industries and plans to extend this facility to the construction industry in the near future under the auspices of the Construction Skills Certification Scheme. The cost of training received via the CTA card varies dependent upon the machine type trained upon and whether the operator claims to have any previous experience or not. For operators without previous plant operation experience, fees include Intermediate Construction Award (ICA) registration, course modules, CTA scheme registration, health and safety test and tuition costs. For those companies who are CITB levy payers, attendance, achievement and 25 per cent supplementary grants are available from the CITB which claim to cover the cost of training for new operatives.

Other schemes, such as the Contractors Mechanical Plant Engineers (CMPE), guarantee competence for plant operation across industrial boundaries. There are no grants available for training and certification, but the CMPE do subsidise such costs by reducing fees to their members. Other equally credible schemes include those run by: National Plant, Lantra and the National Fencing Contractors Association. There are also more specialist card schemes. For instance, the Powered Access Licence issued by the International Powered Access Federation (IPAF). IPAF claims that this pan-European scheme has already trained over 25,000 operators and is the most widely recognised body responsible for improving powered access safety. For specialist forklift trucks the Association of Industrial Truck Trainers (AITT) provide expert training on forklift trucks (although they also train to a high standard on other off-highway vehicles). Similarly within the quarries and mineral extraction industries, EPIC is undoubtedly the leading training provider. In many cases, specialist training represents the only viable means by which to train operators and hence, the trend to 'specialise' is probably set to continue. However, a Voluntary Code of Practice (VCOP) for off-highway plant operator training and competence development (originally written by National Plant) has been published. This document represents a major step forward as it will ensure a more homogeneous standard of operator training throughout industry (and will not be restricted to the construction sector).

Differentiating between training and competence

It is crucial to understand the difference between training and competence (and interpretation of these terms) and the potential legal consequences. Ownership of an operator's training card does not 'automatically' mean that the operator is competent and safe. Rather, operator certification ensures that a base standard of training has been received. This general approach of training achievement, as a precursor to competence acquirement, has been broadly adopted because of the inherent difficulties associated with the myriad of bespoke operational environments and machine types that could be encountered. For example, one machine type such as excavators, can be available in a wide range of formats, each with various operational attachments and unique operational and maintenance requirements. The initial and relatively safe training environment will only partly prepare the operator for industrial application. Training provides familiarisation with core operational, health and safety awareness and other essential job knowledge issues. Nevertheless, it would be unreasonable to expect the newly trained operator to be fully proficient without having acquired substantial industrial experience as well.

To breach the gap between training and competence, industry and government (within the UK and Australia) have co-developed 'competency based' vocational qualifications for

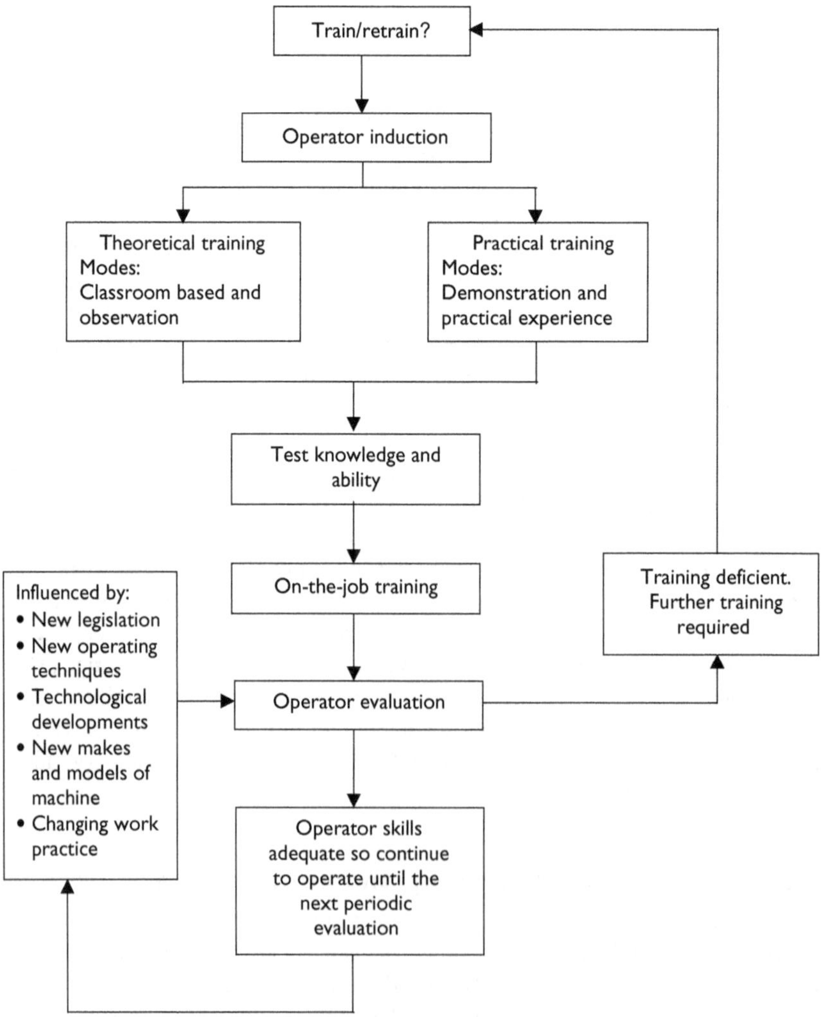

Figure 11.2 The route to competence.

plant operation. This has been achieved in the UK, for example, by the alignment of the CTA card to the Scottish/National Vocational Qualifications (S/NVQ) via the Intermediate Construction Certificate (ICC). To date, initial subscriptions to the S/NVQs have been slow but demand is probably set to increase as professional and industrial bodies and government exert further positive pressure upon practitioners and training providers.

The route to competence

Regardless of the training approach employed or card attained, the competency of an operator at performing specific tasks on specific plant items requires employer time and finance, and operator devotion. There are three fundamental stages that underpin operator competency. These are:

1. *Formal training*, to include company induction, theoretical training and practical training. This essentially aims to provide operators with instructions regards good industrial procedures and practice and will be tested to ensure that required standards are achieved.
2. *'On-the-job' training experience* which commences the process of practical skills knowledge and operational ability acquisition.
3. *Continual employee evaluation* which will ensure that operators retain core skills and attain new ones thus maintaining competency skills. Refer to Figure 11.2.

A combination of these three components will ensure that the trainee receives and comprehends instructions given and utilises the information to safely operate the plant item in an efficient and effective manner. Each stage will now be explained in more detail.

Formal training

Formal training encompasses company induction, theoretical and practical training and testing knowledge retained. Company induction introduces the operator to the organisation, culture and way of thinking. In addition, theoretical and practical training will provide the operator with the skills to fulfil basic organisational requirements and objectives. Consequently, business aims have a far greater probability of being realised. Information on training received by the operator is recorded onto an operator training achievement matrix (refer to Figure 11.3). This enables management to determine what training has been undertaken and areas where further training is required. Such a matrix also charts whether the operator has acquired sufficient operational skills.

Company induction

The company induction process informs operators of their roles and responsibilities within the organisation, to themselves, and to others. The process commences with an introduction to the new work environment. Ideally, this induction should start on the first day of the operator's employment. An information pack, presented to the operator, will help the new

D.J. PLANT LIMITED (Est. 2002)
Training achievement matrix
(Records maintained by Jo Bradley, Director of the Service Department)

Name	Job title/ Position	Safety awareness	LOLER awareness	PUWER awareness	Abrasive wheels	COSHH awareness	Plant operation training certification	Health and safety training certification	First aid	Health and Safety Test	Fire extinguisher awareness	Compressed gas safety awareness	Electrical awareness	General industrial awareness	Periodic maintenance and oil sampling techniques	Banksman and slinger interaction/operation	Service parts ordering and purchasing	Minor structural repairs and welding	Machine specification details	Familiarisation with plant operations manual	Fuel storage and contaminant prevention
F. May	Operator	**	**	**	**	**	**	**	**	**	**	**	**	**	**	**	**	**	**	**	**
G. Carter	Operator	**	**	**	**	**	**	**	**	**	**	**	**	**	**	**	**	**	**	**	**
J. Williams	Service	**	**	**	**	**	*	*	*	*	*	*	*	*	*	*	*	*	*	*	*
J. Nicholas	Operator								*	*	*	*	*	*	*	*	*	*	*	*	*
L. Holt	Service	**	**	**	**	**	**	**	**	**	**	**	**	**	**	**	**	**	**	**	**
M. Trevis	Service	**	**	**	**	**	**	**	**	**	**	**	**	**	**	**	**	**	**	**	**
P. Daniel	Foreman	*	*	*	*	*	*	*	*	*	*	*	*	*	*	*	*	*	*	*	*
R. Jones	Operator								*	*	*	*	*	*	*	*	*	*	*	*	*
T. Wood	Operator	**	**	**	**	**	**	**	*	*	**	**	**	**	**	**	**	**	**	**	**

Key: * denotes operator training received; and ** denotes operator competence.

Form No.: DJP126 | Issue No.: 1 | Issue date.: 29/09/03

Figure 11.3 Training achievement matrix.

employee digest relevant information. Material covered in an induction pack should include the following:

1 an overview of the organisation's structure, business objectives and current activities;
2 a summary of the position of the operator within the organisation with respect to the previous point;
3 a job description together with specific duties;
4 the conditions of work, including wage structure, holiday entitlement, grievance procedure and so on;
5 an introduction to the training and development plan;
6 a copy of the organisation's health and safety policy;
7 a description of plant item(s) to be operated.

Company induction serves several functions. It provides immediate and accurate information that helps the operator understand essential employment duties and responsibilities. It also enables the operator to 'feel part of the organisation'. Induction therefore provides the new operator with an optimum start to their employment and relieves unnecessary anxiety.

Theoretical training

The purpose of 'taught' plant theory is to instil operators with relevant information that will ensure safe and productive plant operation. Operators are therefore forewarned of any potential hazardous situations, their potential consequences and how to avoid them.

Source material

When teaching theory, the training provider can refer to a wide range of written material. For example, the machine operations and maintenance manual, current health and safety legislation, codes of practice, company and job specific site rules, etc. These documents must be read and understood by both managers and operators as part of the process and prior to the operator commencing training on the plant item itself. In a practical context an experienced operator attending such a course may well lack theoretical knowledge pertaining to the plant item they currently operate. Because an operator has operated a machine for many years, it does not necessarily follow that they have operated the machine correctly, or to its best ability. Experience of plant operation itself is not enough to ensure that operating techniques employed are efficient and safe.

Method of delivery

The method utilised for delivering theory can be varied and largely dependent upon the training provider chosen. Nevertheless, it should be recognised that each operator has individual needs and that their rate of learning can be increased (or otherwise) by the application of different stimuli. Some operators can best learn instructional material independently while seated in a classroom, others require a more nurtured and supervised approach and some benefit more from direct participation. In most cases, a combination of different stimuli delivers the optimum level of learning in the least amount of time.

Formal education need not take place in a classroom. Discussions can consist of the trainer verbally explaining the educational material to the trainee, either in the workplace or on site. Alternatively, the Internet is an expanding, versatile and reliable medium through which to educate, train and test plant operatives. The operator and plant manager should jointly decide upon the most suitable method.

Mechanisms for delivery

Plant training must include an explanatory 'theoretical' element as well as a practical element since this combination ensures greater knowledge retention. A variety of formats may be employed as part of formal instruction. Such may include lectures, conferences, classroom discussions, demonstrations and written or oral tests. The use of visual and audio aids, for example, films, slides, computers, videotapes and other visual presentations, may also enhance the training provided and facilitate the learning process. Visual aids have certain advantages over formal lectures and demonstrations:

1 Trainees are likely to remain more attentive, thereby increasing the training's effectiveness as well as stimulating constructive debate between all parties.
2 Training providers can convey the necessary information in a reduced amount of time.
3 Trainees are more likely to acquire understanding than through an extensive non visual description.
4 Trainees will have greater retention of information.

Once the method of delivering the theoretical aspect of the training has been selected and implemented, consideration should focus upon practical training on the machine itself.

Practical training

The practical element of a training course should consist of a combination of observation and supervised 'hands-on' operation, that is, the trainee plant operator will receive a demonstration of basic machine manoeuvres and maintenance performed by a competent operative (trainer) prior to acquiring practical experience. The practice area should be as level and clean as possible so as to avoid any unnecessary danger. This will allow the operator to acquire operational confidence and basic operating skills. Thereafter, the training should aim to simulate a more realistic work environment whereby the operator is gradually introduced to various site conditions of increasing difficulty and alerted to potential hazardous operational conditions. At this stage, the trainee can begin learning more complicated manoeuvres and techniques.

Consideration should also be given to the machine make and model that the operator is trained upon as each has different operating characteristics, limitations and other unique features. For example, consider the new Caterpillar 432D wheeled backhoe loader that includes innovative pilot controls to enable ease of operation when compared to traditional lever controls. 'New' operators receiving taught instruction on these machines may well find some difficulty operating 'traditional' models that still exist (and are still manufactured). In addition, some machines can be fitted with specialist attachments that may affect equipment operational techniques and procedures. Therefore, the ideal training program should be based upon the type of vehicle that the trainee would be authorised to operate.

Table 11.1 represents an outline of a generic operator training program in which characteristics of the plant item(s) that the operator will operate are obtained.

The operator should also always refer to the manufacturer's operations and maintenance manual that accompanies the individual machine operated.

Table 11.1 Essential content of an operator training program

Item	*Description*
Machine anatomy	Machine anatomy on a compartment level (i.e. engine, transmission, hydraulics, etc.) with some reference to individual core components (i.e. filters, hydraulic hose 'O' rings and so forth) Machine size restrictions Machine visibility to include: manual, mirrors and rear-view cameras (if fitted) Vehicle capacity and stability
Machine maintenance	Daily maintenance procedures with operator roles and responsibilities Machine refuelling and battery changing and/or recharging Extracting oil samples for analysis Record keeping
Machine operation	Machine start-up procedures, vehicle inspection and machine shut down Machine steering, manoeuvring Controls, instrumentation and any display panel with particular emphasis on their location, what they do, and how they work Machine attachments, their functionality, operation and any limitation of their use Typical operating applications and limitations Operating machines on ramps, sloped surfaces or precarious ground conditions that might affect vehicle stability or at least increase operational difficulty Operating procedures when working in narrow transportation routes/aisles and/or areas with restricted space limitations
Machine health and safety	Awareness of site safety 'visual' signs and banksman/foreman hand signals. This will warn operators of hazards and provide suitable guidance on either safe working practices or correct vehicle routes. All signs should comply with the Safety Signs and Signals Regulations 1996 Awareness of roll-over protective structures (ROPS) that are fitted to vehicles where the risk of over turning is significant or falling-object protective structures (FOPS) where there is a significant risk of falling materials, for example, demolition Awareness of on-board visual and audible alarms and how to interpret them Industrial knowledge and awareness of the operating environment, that is, construction, mining, demolition, etc. Any other operating instruction, warning, or precaution listed in the operator's manual for the type of vehicle the employee is being trained to operate

Practical and theoretical test

Upon completion of the initial training programme, the training provider must finally evaluate the trainee. Two fundamental aspects of the operator must be tested, namely, knowledge retained and ability acquired.

Testing knowledge retained

To determine whether the operator has retained adequate knowledge pertaining to the machine and its operational environments, a test is required. A significant proportion of the questions should be plant specific, that is, appropriate to the plant item to be operated. These may include topics concerning fuel storage, mounting and dismounting the machine and so on. The remainder of the questions will be on more general topics, usually presented in a multi-choice format. For example, questions on who is responsible for managing health and safety on site or conducting risk assessments are typical. The test should be delivered through an appropriate medium, dependant upon the individual trainee. Some operators may be more comfortable with a written test; others with a computer-based test. Where operators lack basic written skills or feel intimidated by these methods, a verbal examination may be more appropriate. The training provider should not confuse academic qualification with intelligence. Whatever the means of testing, the standard of questions must be high enough to ensure that the trainee contributes to a safe, productive working environment and should be relevant to the operator's daily tasks and duties.

Testing ability acquired

The operational ability obtained by the operator through the practical element of training received must also be tested. A predetermined benchmark standard for each item of practical training is retained by the assessor and is used as a basis to compare the operator's actual performance against during a field test. All instruction should be conducted by a person with the necessary knowledge, training and experience to train operators and evaluate their competency. This may be the employer, another employee or other qualified person if attending an external course. The training and evaluation does not have to be conducted by a single individual, but can be done by several persons, provided each one is suitably qualified.

Test security

Test administration can be difficult in the absence of rigorous systems and procedures that ensure that the operator who receives training is the same operator that attains accreditation. There have been cases in the past where operators have fraudulently obtained operator certification. To avoid certification registration abuse, thorough and meticulous documentation is required. Figure 11.4 represents a template record of a 'training received' form. The first part of the form records personal details of the operator along with a photograph for security purposes; this is to ensure that the person who attains the card is the same person who has undertaken the training. The second part of the form will detail the type and model of machine operated together with a description of training received and signatures of both employer and employee.

Ideally, the training package should be divided into core modules and upon completion of each module, a record of training received should be completed and retained by management.

On-the-job training

On-the-job training represents the final stage of a comprehensive training programme as it introduces the trainee 'journeyman' operator to the work environment. During this period, the operator must be carefully monitored in order to ensure a continuation of best practice. A probation period of anywhere between several months and several years is advisable, dependent upon the machine type and operational environment.

D.J. PLANT LIMITED (Est. 2002)
55 Clent View, Dudley, West Midlands, DY4 5ZZ
Tel: 01384 546978-000, Fax: 01384 546978-001, Email: djp@plant.com
TRAINING RECORDS

Employee Personal Details	Record Details
Name:	Employment date:
Age:	Form No.: DJP125
Job Title/Position:	Issue No.: 1
Address:	Date of Issue: 29/09/03
Telephone No.:	
Email:	

Date	Machine/ Model	Description of training received	Employee signature	Employer signature

Figure 11.4 Record of training received.

Specifically, on-the-job training should emphasise the operating environment together with any peculiar features, as these will affect vehicle operation and maintenance practice. A more arduous operational environment may merit the use of a more rigorous daily maintenance regime. For example, quarrying and mining is perhaps one of the more erosive environments upon a machine, therefore particular attention should be paid to ground engagement gear wear as well as seal and filter deterioration. Consideration should also be given to the function of plant and equipment and how the machinery interacts with personnel on site. Other factors which should receive due consideration include:

1 Familiarisation with pedestrian and vehicle traffic routes in order to ensure, as far as is reasonably practicable, that risks to the health and welfare of operatives are minimised. The operator will also have to be aware of floor surfaces and/or ground conditions where the vehicle will be operated since this can influence machine operation.

©The Contractors Mechanical Plant Engineers

Plant Operators Certificate of Competence – Record of Competence

Plant Operator's Details: Book No:................ Page No. 1

Name: .... CMPE Certificate of Competence No: ....

Employer's Name: ....

Tel No: .... Fax No: ....

Type of Plant Operated: ....

....

Type of Work Carried Out: ....

....

....

Customer Details

Name: ....

Head Office or

Regional Office: Tel No: .... Fax No: ....

Site Details Name: ....

Address: ....

.... Tel No: ....

Work on Site from: .... **to:** ....

Comments: ....

....

....

....

Please Print:

Name: .... Signature: ....

Position .... Date: ....

Thank you for taking the time to complete this page it will assist the operator in the future.

Figure 11.5 CMPE operator's logbook.

2 For vehicles that carry loads, the composition of probable loads and determination of load stability. Similarly, an in-depth knowledge of load manipulation, stacking, unstacking, loading or dumping procedures should be acquired.
3 Operating techniques, systems and procedures when working in classified hazardous locations that exist or may exist in the workplace.
4 Operating the vehicle in closed 'unventilated' environments, trenches, or confined spaces where insufficient ventilation and/or poor vehicle maintenance could cause a build-up of noxious carbon monoxide exhaust fumes.

Upon completion of the probationary period, the employer should then complete a certification of training record which collates core information on the name of the operator, the date of the training, the date of the evaluation and the identity of the person(s) performing the training or evaluation. This information is retained in an operator logbook which can be readily obtained by both the operative and the employer, or alternatively in the case of an accident, the Health and Safety Inspector. The employer must certify that each operator has been trained and evaluated as required by the organisation's standards. Figure 11.5 provides an example of an operator logbook issued by the CMPE. Each page of the logbook records the core information listed above and consists of five main sections; namely operator details, plant specific details, customer details, site details and comments. Within the CMPE logbook there are numerous blank pages that allow for a new form to be completed at the end of each working week.

Potential future developments include the introduction of a smart card (similar to a credit card) that contains essential information on the operator's training and experience. Only when this information is at the required standard should the employer, or training provider, issue the operator with a certificate or card of competence.

Employee evaluation

The plant and equipment sector is subject to continual development. Competency acquirement and the efficient operation of plant were earlier identified (Figure 11.2) as part of a dynamic process, one which can be influenced by new legislation, operating techniques, technological developments, makes and model of machine and changing work practice. The operator's core competence may remain constant but general knowledge may from time-to-time require 'top-up'. It is therefore not sufficient to merely acquire knowledge and skills. The better organisations clearly recognise the continual need to re-educate and develop operators in line with current developments within industry.

When the operator completes the training exercises and initial approbatory on-the-job operational experience, an evaluation of the employee must be performed. Essentially, this evaluation will determine the adequacy of training and the ability of the employee to perform plant operations safely in the workplace. Most credible training programs ensure that the trainee operator is evaluated on a cyclical basis, often every 4–6 years. Primarily, the evaluation will seek to determine whether the operator has retained sufficient skills and competence to continue operating the equipment.

An evaluation may also be required where the site manager observes the employee performing an unsafe act, such as travelling at an unsafe speed. The person making the observation should point out the incorrect manner of plant operation or other unsafe act being

D.J. PLANT LIMITED (Est. 2002)
55 Clent View, Dudley, West Midlands, DY4 5ZZ
Tel: 01384 546978-000, Fax: 01384 546978-001, Email: djp@plant.com

Employee name	Area where further training/instruction is required	Actions taken to address areas of deficiency	Instructor name and details	Programmed date of instruction

Form No.: DJP127 | Issue No.: 1 | Issue date.: 29/09/03

Figure 11.6 Training monitoring.

conducted, tell the employee how to do the operation correctly, and then ensure that the employee does the operation correctly. All information should be recorded on a training monitoring report form as illustrated in Figure 11.6.

When there have been multiple on-the-spot 'corrections', the employer may decide to conduct a more structured programme of refresher training or even, in the worse case scenario, retrain the operator. Refresher training is required when the operator has been observed driving unsafely, has been involved in an accident or near-miss, has received an evaluation that indicates unsafe operation, is assigned to drive a different machine, or if a workplace condition affecting safe operation changes. The training would include the following information:

1. common unsafe situations encountered in the workplace;
2. unsafe operating methods observed or known to be used;
3. the need for constant attentiveness to the vehicle, the workplace conditions and the manner in which the vehicle is operated.

The above subject areas need to be taught so that the operator receives all the information needed for safe vehicle operation. Upon completion of the refresher training, the operator must again be monitored to ensure compliance with organisational and legal requirements.

Chapter 12

Operational planning of equipment as a resource

Introduction

> The finest of plans are always ruined by the littleness of those who ought to carry them out, for the Emperors can actually do nothing.
>
> Mutter Courage (1939)

A plan is defined in the Concise Oxford Dictionary as 'a formulated and especially detailed method by which a thing is to be done'. What this simplistic definition does not disclose is that to be effective, there must be commitment to the plan from everyone who is affected by it. History is littered with some of the finest leaders in warfare and business that developed the most intricate and accurate plans only to have them fail because of a lack of workforce commitment. The converse is also true since a 'highly skilled' workforce needs good leadership and effective planning to realise its maximum potential.

Planning for future business activities exploits historical, current and forecasted performance information to provide guidance for an organisation, its individual projects and perhaps more importantly, the employees within it. Plans are formulated in advance and provide a unifying framework of decision making throughout the organisation. This unified 'sense of direction' is of particular relevance since the absence of a plan exposes an organisation to unnecessary 'chance' and possible loss of business revenue/profits.

Planning is undertaken at different levels within an organisation and can focus upon either strategic issues (i.e. formulating the company's future aims and objectives) or operational issues (devising appropriate processes and procedures to achieve these aims and objectives). Operational planning therefore seeks to improve the efficiency and effectiveness of the services and products provided by the business to its clients. Essentially, operational planning helps departmental managers and employees convert strategic plans into detailed monthly, weekly and daily tasks and objectives. Within this chapter the application of operational planning to the management of plant and equipment is presented along with techniques that could be utilised. These include data management, statistical process analyses, plan representation and subsequent dissemination throughout the organisation.

Operational planning process

Operational planning commits the resources and actions required to successfully achieve organisational objectives. The 'plan' can be as exhaustive as required but the degree of detail will ultimately depend upon the nature of the business and its plant holding. Larger companies with huge financial investment in plant and equipment (and ancillary services, e.g. equipment inventory storage, management software, mechanical fitters and so forth) would require more meticulous plans than say the owner operator. Nonetheless, planning, for any plant owner, can make the difference between a successful and failed business venture.

Prior to plan preparation, management should consider and discuss what types of off-highway plant management strategy are to be employed, both within the organisation and per project. All relevant factors should be considered such as finance arrangements, production requirements, health and safety, maintenance management and human resource management. Such information can be obtained from data held within the company itself or sought from external consultants. After careful review, an appropriate operational plan should be prepared to ensure that enhanced plant performance is achieved.

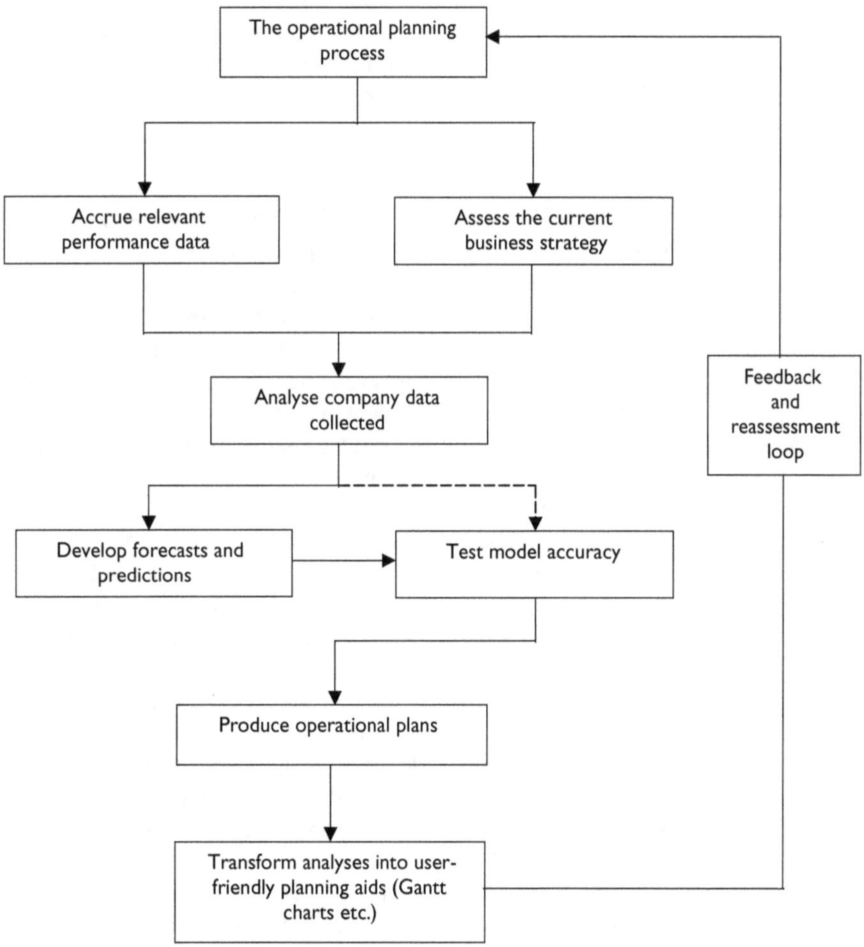

Figure 12.1 The operational planning process.

Planning is not a singular activity but rather a process of activities that culminates in the production of a plan and its subsequent implementation. However, planning should be an iterative process, the plan requiring periodic assessment during and post implementation (using a control loop). Figure 12.1 represents the process leading to the development of an operational plan.

The first step in the planning process is to conduct an overview of the organisation's operational activities in order to assess the current business strategy. Second, relevant organisational data must be collected and analysed to determine the existing performance of plant and equipment under various operational scenarios. Such data will help to establish performance benchmarks and enable the future production of more accurate operational plans. Performance data gathered can then be modelled using one of several deterministic techniques (e.g. time series analysis or regression). For example, the travel speed of a dump truck is dependent upon many variables, such as ground conditions, whether the machine is loaded or not, the angle of slopes, the specifications of the machine (horsepower etc.) and so on. Accurate modelling of these variables enables the impact of each to be determined such that future estimates of machine production (and thus profit), for any given operational scenario are reliable. Output from the forecasts and predictions is then assessed to prove the robustness and accuracy of the models. Once proven, the next step is to transform the models developed into user-friendly planning aids, such as Gantt charts. Finally, the plans and models are periodically evaluated and improved.

Information management and plant performance study

Crucial to the planning process is the accumulation and manipulation of business information and plant performance data. Information acts as a conduit that facilitates a manager's understanding of a plant organisation's interaction with competitors and clients. Documentation is therefore a valuable resource comparable in value to the traditional resources of capital, labour and materials. Records relating to plant items are no exception since comprehensive records of plant purchase, maintenance and operation have the capacity to generate a more 'streamlined' cost effective strategy and acquire a competitive advantage. Accurate records of machine performance enable management to develop optimum procedures and practice. Plant organisations should endeavour to determine which items are the most profitable, how profits compare across various projects, the size and movement of inventory store items, staff resources and whether these are at an optimum level and so forth.

To capitalise upon information collected, sophisticated management information systems (MIS) have evolved to assist managers in their primary tasks and organisations in conducting their business. MIS represents the systematic control of information such that organisational aims and objectives are met; it is therefore a tool that is exploited for the benefit of a company. Most organisations have an MIS; this can be manual, computer based or a combination of the two. Computer based systems do offer greater advantages in terms of accuracy, efficiency and data manipulation. Regardless of the method employed, the underlying objective of an MIS is to provide useful, factual and appropriate information when required and not just spurious reams of data with no real use or meaning.

Those contemplating the development of an MIS will first need to identify the type of data required and determine how such could be collated. Two types of data are available, namely secondary and primary data. Manufacturer specifications and applications handbooks (particularly those developed and published in performance handbooks, e.g. by

Caterpillar and Komatsu) are a readily accessible source and provide a wide variety of meticulous 'secondary' information regards machine costs, expected performance and longevity under given scenarios. Whilst this information can form the basis of a plant management strategy, it may contain some bias and may not provide a complete solution. Therefore, when predicting production output, financial expenditure or when using the production output to develop plans, the collation and analysis of 'primary' information is important.

Work environment, operator skills and competence, machine activity (for instance, breakers attached to excavators) and so forth, all impact upon machine performance and therefore must be measured. Work-study provides an opportunity to measure these primary data sources and can be described as 'the systematic examination of activities in order to improve the effective use of human and other material resources' (Pilcher, 1992). Thus, work-study collects primary information and aims to acquire a greater understanding of operational processes and procedures so that subsequent refinement and improvement can be made. It represents a critical self-appraisal of the work environment and facilitates effective streamlining of the company and its plant holding to be made.

A typical example of the potential benefits to be accrued from using work-study is provided by considering a practical problem where a backhoe 360° tracked excavator has been loading overburden into a rigid dump truck. Management complained to the manufacturer that predicted production rates had been substantially lower than actual rates. Upon work-study examination of the loading activity, it became apparent that the excavator was situated on the same substrata level as the dump truck. This meant that the excavator had to load the bucket, slew to the dump target, and importantly, raise the machine boom, dipper and thus bucket, before discharging the overburden. The position of the machine, in terms of height relative to the dump truck, meant that the machine cycle time was increased significantly as a result of the machine arm having to be raised to the depth of the excavation plus the height of the dump truck. After further consideration, a mound of earth was constructed to act as an operating platform (otherwise known as a bench). This raised the height of the machine above the dump truck and reduced the cycle time dramatically. Essentially, this meant that once the machine had filled its bucket and was at the top of the bench it was also at the correct height to slew to the dump target, thus reducing the 'lift' time. The emphasis in this example is on machine production but the process of reviewing, assessing, developing a new process, recording data (in this case cycle time) and measuring whether the new process is an improvement upon the previous can be applied to many other plant management problems.

Data analysis techniques

Once the data (historical, primary and secondary) has been collected, business forecasts can be developed. Forecasting is an integral part of operational planning. When managers formulate a plan, they strive to determine in the present what necessary courses of action their organisation should take in the future to ensure a successful business venture. On occasion, unpredictable events will significantly influence the effectiveness of the forecast and render it less accurate. Indeed, there are so many unpredictable social, economic and political variables that sceptics would suggest that any prediction is futile. Nonetheless, some assumptions about the future are more reliable and include: the availability of financial, human and material resources; changes in mechanical (and other) technology; competitor strengths and weaknesses; and government taxation. Even when using these relatively stable statistics, any prediction made will never be 100 per cent accurate. However, it is far better to supplement 'gut feeling' of a trend with a rational assessment of it – forecasting is both an art and a science!

Essential ingredients of a good forecast are that it is easy to use and interpret and is practical. Complexity is not always the best solution; some of the most sophisticated economic forecasts have been a failure. Unfortunately, some managers base reports and plans upon simple trends and intuition. Thereafter, an inaccurate interpretation of the data is made, using a mixture of subjective hunches and aspirations that are grounded neither in today's realities nor tomorrow's scientifically projected future. Data is undoubtedly an important ingredient in preparing effective plans but it requires insightful manipulation and consideration to be useful.

The following techniques represent an introduction to fundamental deterministic quantitative statistics that have already been successfully applied within off-highway plant management research work. It is not the intention to prescribe a complete data analysis strategy that would naturally achieve effective management of plant. Rather, examples given will assist students and practitioners develop a fundamental understanding of how data can be successfully exploited in practice. Whilst the information contained herein is by no means definitive, it will act as a catalyst to learning more complex and novel techniques.

Some a priori knowledge of basic statistics has been assumed together with a rudimentary knowledge of Microsoft Excel. The analysis will utilise Microsoft Excel as this is a popular, widely used and thus, accessible package amongst students and practitioners; albeit, other analysis packages such as Statistics for the Social Scientist (SPSS) and Minitab are worthy of consideration. The following text has been designed to be read quickly and absorbed with a minimum amount of effort. To reinforce the learning process, a series of worked examples, complemented with appropriate screen dumps, has been provided. The reader is urged to work through these examples and follow the step-by-step process.

Correlation (or statistical association)

Correlation analysis examines and explores the relationship between two (or more) variables to measure the strength of association between them. Consider the travel time of a rigid dump truck and the distance it travels. If these two variables are plotted they could form one of three scatterplots; illustrating some, perfect or no association (Figures 12.2–12.4 respectively).

Although the trend for some and perfect association shown in these examples is linear (i.e. a straight line), the association may alternatively be curve linear (e.g. exponential or cubic). It is therefore advisable that a graphical representation of the data be examined before continuing with any further analysis.

Pearson's product moment correlation coefficient

Although the scatterplot can provide some indication of the correlation present between two variables, measuring the actual amount of correlation present can be achieved through the use of Pearson's product moment correlation coefficient (r). To use r, the data must satisfy two criteria. First, data should be measured on an interval or ratio scale, both of which are 'continuous' data and represent scales of measurement such as time, heat and weight.

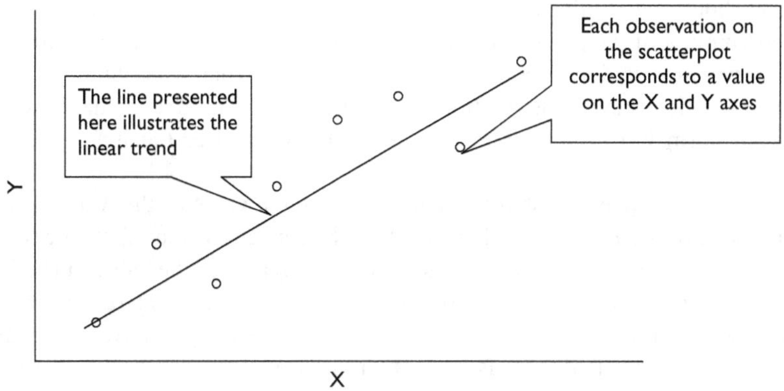

Figure 12.2 Some association.

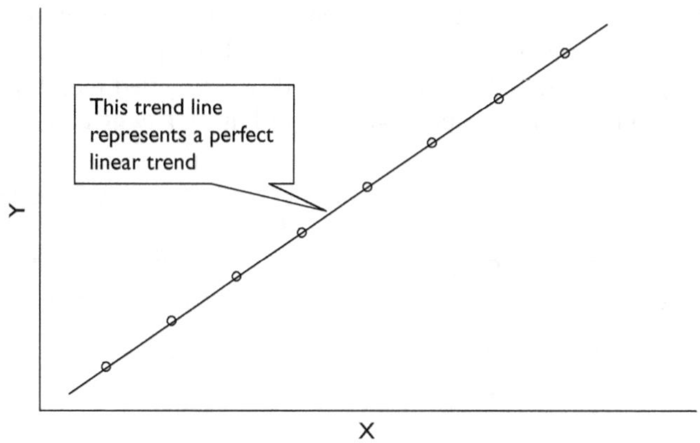

Figure 12.3 Perfect association.

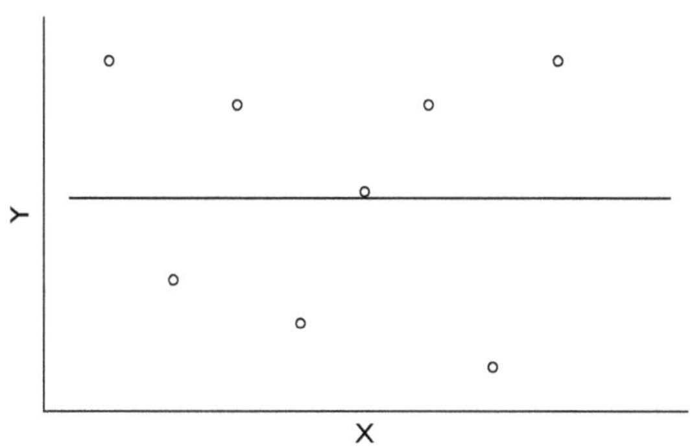

Figure 12.4 No association.

Kvanli *et al.* (2000) provide comprehensive details of these and other measurement scales. Second, the data should satisfy the assumption of normality. Normality is represented by a symmetrical distribution of the given data such that the mean, median and mode are equal; typically, this distribution is described as being 'bell shaped'. Those distributions that are not bell shaped are called skewed distributions and therefore non-normal. In the absence of the assumption of normality, a non-parametric test should be utilised (such as Spearman's correlation).

If two variables are positively related to each other then the value of the X axis variable increases as the value of the Y axis variable increases. The converse is therefore true for negative correlation, that is, the value of the X axis variable decreases as the value of the Y axis increases. The value of Pearson's correlation coefficient (denoted as r) ranges in value between 1 (for perfect positive correlation) and -1 (for perfect negative correlation). A value of 0 therefore means that there is no correlation. Because r is a coefficient it does not have any units of measurement.

Example 12.1

A site engineer knows that (as a general rule) the rate of site productivity shares a linear relationship with the actual number of 'working hours' each week. That is, longer working hours equate to higher productivity. A series of production data for plant operators operating backhoe loaders (specifically, m^3 excavated per hours worked) are recorded monthly over a 12-month period (refer to Figure 12.5). Ten operatives are working at any given time.

Microsoft Excel - Correlation

C3 = 81250

	A	B	C	D
1		Month	Machine output (m3)	Hours worked
2	January	1	75000	130
3	February	2	81250	130
4	March	3	87500	140
5	April	4	112500	175
6	May	5	100000	155
7	June	6	120000	178
8	July	7	120000	175
9	August	8	118750	185
10	September	9	90000	155
11	October	10	87500	140
12	November	11	81250	130
13	December	12	75000	135

Figure 12.5 Production data.

Data examination

The first step is to view the data for any trends that may be apparent. An initial perusal of the data identifies that the working hours seem to be lower during the winter periods and increase as climatic conditions improve. One would expect to find such a trend since inclement weather (rain and frost) can hinder or even halt production work.

Scatterplot production

The second step is to construct a scatterplot in order to view the pattern between the hours worked (*Y* axis) and the machine output (m^3) (*X* axis). To do this, highlight cells 'C2:D13' and click on the `Chart Wizard icon` (refer to Figure 12.6). Select `XY (Scatter)` and then press the `Next` button and cycle through a series of windows to produce the graph illustrated in Figure 12.7.

Figure 12.7 provides evidence that a positive linear relationship exists between machine output and hours worked. However, this is purely an observation and somewhat subjective, therefore further study is required.

Correlation

To calculate *r*, click on the `Tools` drop down menu and select `Data Analysis`; highlight `Correlation` and then click `OK`. In the `Input Y Range` insert 'C2:D13'. The output to the analysis is provided in Figure 12.8.

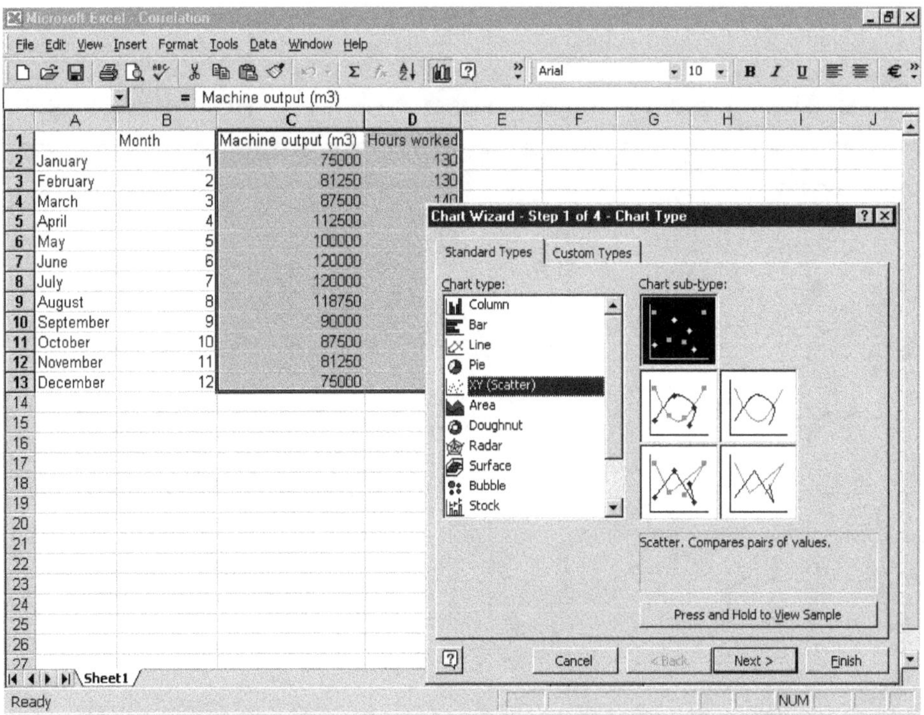

Figure 12.6 Chart Wizard.

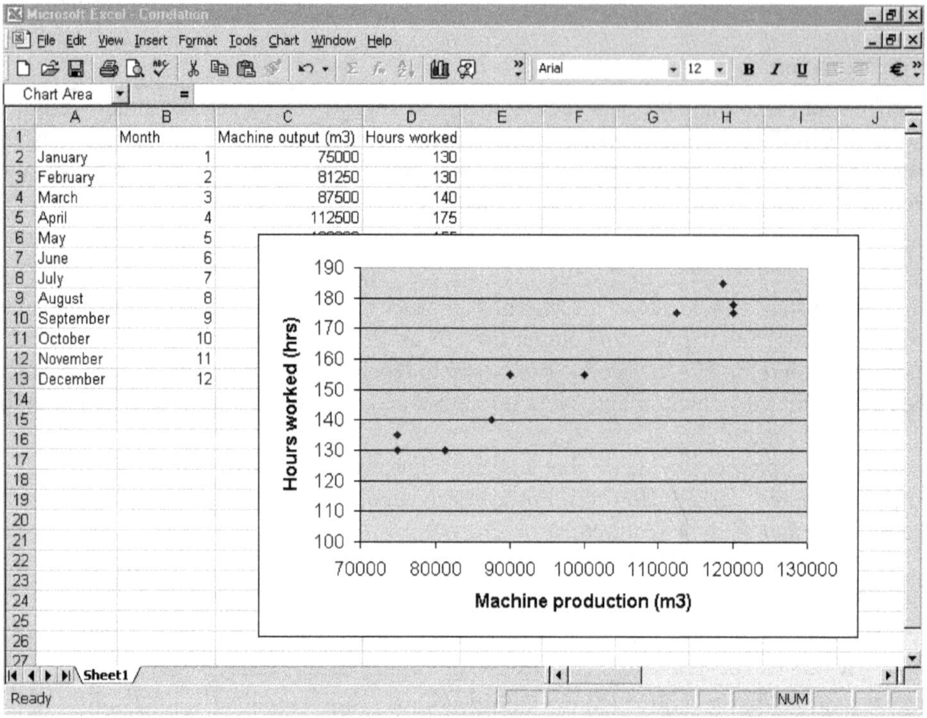

Figure 12.7 Scatterplot of machine output per hour of production.

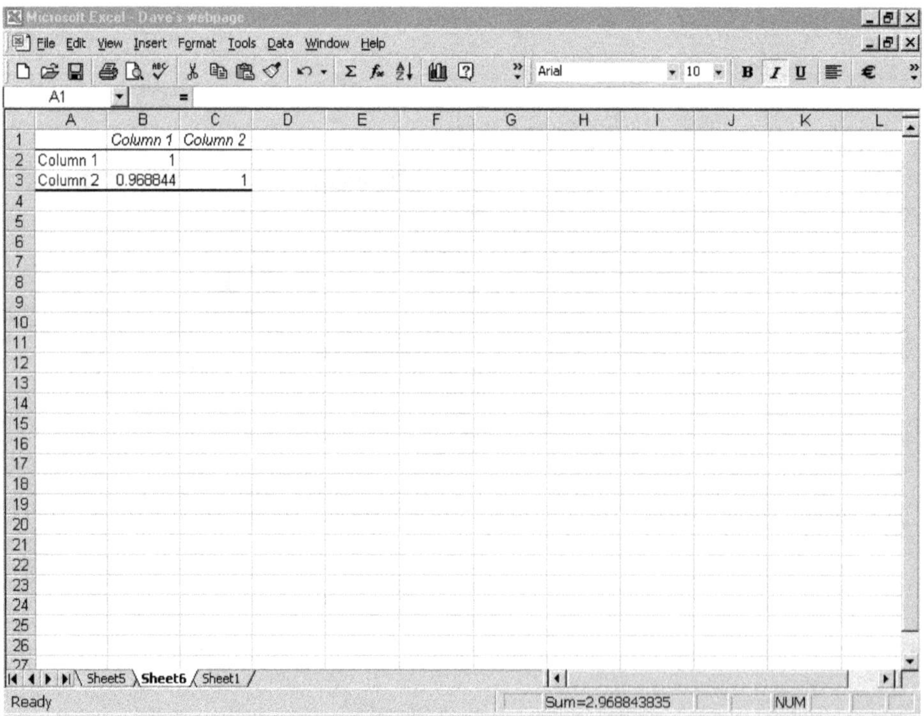

Figure 12.8 Correlation analysis output (*r* value).

The r statistic is located in `Cell B3` and equals 0.968844 (or 0.97 when rounded to two decimal places (dp.)). It is apparent therefore that a very strong positive correlation exists between the two variables and that one variable could be used to predict the other. Note that `Column 1` has a perfect correlation with `Column 1` and similarly `Column 2` has a perfect correlation with `Column 2`. This is because the analysis calculates the correlation of a variable with itself as well as with that of another variable.

It is important to be reminded of the old adage 'lies, damn lies and statistics!' This is because spurious relationships can be found between two completely unrelated variables, for instance, there may apparently be a strong correlation between tin food sales and construction output. It is therefore advisable to ensure that a logical thought process is followed before attempting the analysis similar to the above example of the site engineer.

Linear regression analysis

Having determined that one variable predicts another, the next phase of work could be to develop an equation (model) to represent the relationship between the two variables and make a prediction. Linear regression analysis is a popular method employed to achieve this and simply uses one predictor, the independent variable (x), to predict values of another dependent variable (y). There are two popular types of regression analysis commonly used; these are, linear bivariate regression and multivariate regression. Both methods can be learnt with limited a priori knowledge.

Bivariate regression analysis

The objective of regression analysis is to provide the best fit to the available data; this is achieved using the least squares method (LSM) (refer to Figure 12.9).

Least squares method is used to predict values for a and b such that the straight line regression equation that passes through the observations has a minimum error. Specifically, error is defined as the difference between the predicted values and the actual values of the dependent variable. The basic bivariate regression equation is expressed as:

$$y = a + bx$$

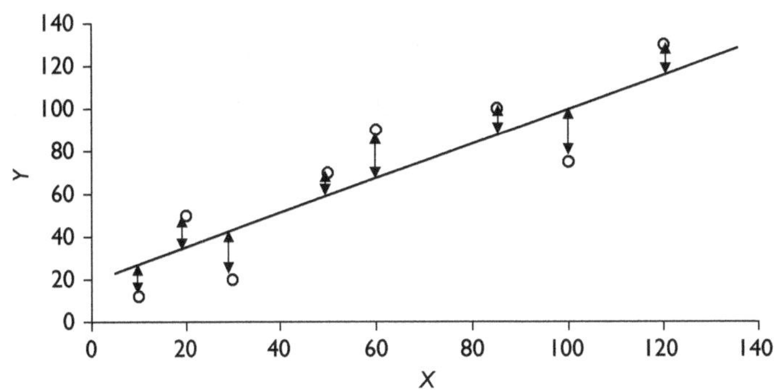

Figure 12.9 Least squares method.

where

- y is the dependent variable (i.e. the one that requires prediction); in the previous example this was machine production output.
- a is the y axis intercept.
- b is a coefficient that represents the gradient of the line that best fits the given data.
- x is the independent variable (i.e. the predictor); in the previous example this was hours worked.

Example 12.2

Using the production data from Figure 12.5 the regression model equation can be determined. To conduct the analysis click on the `Tools` drop down menu and select `Data Analysis` and then `Regression`, then click `OK`. In the `Input Y Range` (machine output; the dependent variable) insert 'C2:C13' and in the `Input X Range` (hours worked; the independent variable) insert 'D2:D13'. Also ensure that the `Residuals` box is ticked before clicking `OK` as this will then allow calculation of both the predicted values for the series as well as the difference between actual and predicted (the residual). Data output is presented in Figure 12.10.

The resulting equation (resulting from cells B17 and B18 'coefficients') is expressed as:

$$y = -28{,}590.5193 + (816.1029 \times \text{hours worked})$$

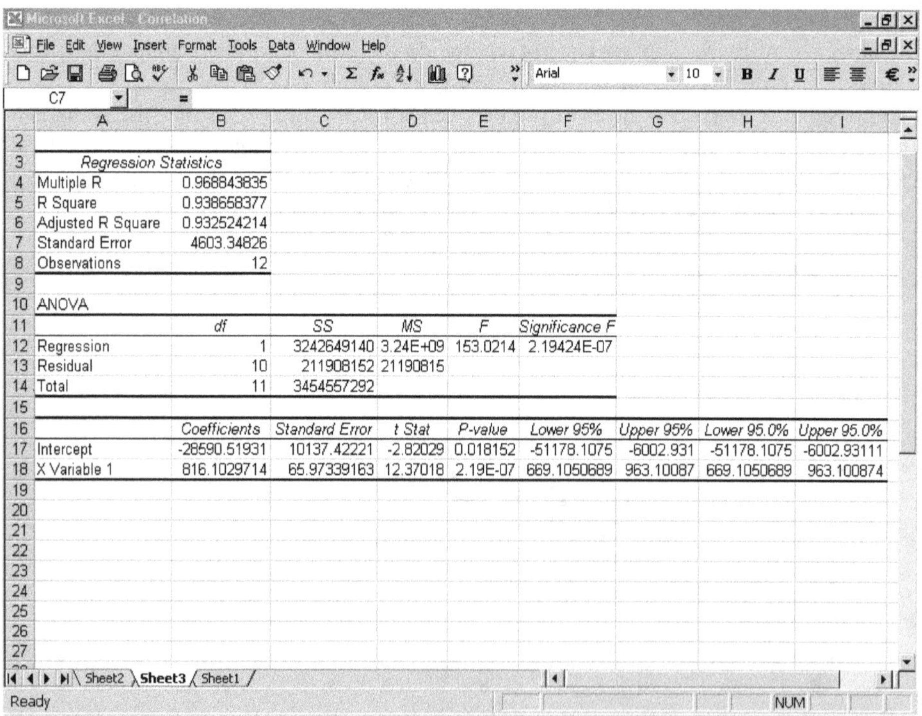

Regression Statistics	
Multiple R	0.968843835
R Square	0.938658377
Adjusted R Square	0.932524214
Standard Error	4603.34826
Observations	12

ANOVA

	df	SS	MS	F	Significance F
Regression	1	3242649140	3.24E+09	153.0214	2.19424E-07
Residual	10	211908152	21190815		
Total	11	3454557292			

	Coefficients	Standard Error	t Stat	P-value	Lower 95%	Upper 95%	Lower 95.0%	Upper 95.0%
Intercept	-28590.51931	10137.42221	-2.82029	0.018152	-51178.1075	-6002.931	-51178.1075	-6002.93111
X Variable 1	816.1029714	65.97339163	12.37018	2.19E-07	669.1050689	963.10087	669.1050689	963.100874

Figure 12.10 Regression analysis output.

This equation needs some additional explanation because the constant coefficient is negative (i.e. −28,590.5193) and therefore, in the event of work stoppage (and thus no hours worked), the prediction would be $-28{,}590.5193\,\mathrm{m}^3$. This is clearly nonsense. This model is therefore limited to making predictions for production data where between 130 and 185 h have been worked (as per the 12 months data utilised). Values either side of these minimum and maximum values would have to be scrutinised in further detail to ensure that predictions are realistic.

To determine whether the model makes reliable predictions of machine production, the `R square` value (cell B5), more commonly abbreviated to r^2, is examined. The r^2 is referred to as the coefficient of determination and represents the square of the correlation coefficient (cell B4). Thus the r value 0.97 'cell B4' squared equals the r^2 value 0.94 'cell B5'. This is an important statistic since it calculates what proportion of the variation in machine output can be predicted by changes in hours worked. Thus, it may be interpreted that 94 per cent of the variation in machine output can be predicted by changes in the value of hours worked.

Multivariate regression analysis

There are circumstances when the strength of correlation between the independent variable and dependent variable is less strong. It may be that there is not just one independent variable but several (i.e. more than one) affecting the dependent variable. In these circumstances a multi-variable approach (using multiple independent variables) is required. Thus, multivariate regression uses more than one variable to predict the dependent variable. The procedure to calculate multivariate regression is purely an extension of bivariate regression.

The basic multivariate regression equation is expressed as:

$$y = a + b_i x_i + \cdots + b_n x_n$$

where

- y is the dependent variable.
- a is the y axis intercept.
- b_i is a coefficient that represents the gradient of the line that best fits the given data.
- x_i is the independent variable.
- b_n and x_n represent the opportunity to add further variables and thus coefficients to the analysis.

Example 12.3

Consider a system previously developed for predicting tracked hydraulic excavator machine cycle time using multiple linear regression (Edwards and Holt, 2000). Three variables were utilised, namely, machine swing angle, machine weight and digging depth (as a percentage of the machine's maximum dig depth capacity). This does not mean that other variables were not important since, clearly, other factors do impact upon machine cycle time (e.g. climatic conditions, type of machine and so forth). Rather the model sought to determine those variables that produced the best prediction with the minimum amount of computational effort.

Cycle time (sec)	Swing angle (degrees)	Machine weight (tonne)	Digging depth (%)
7	45	6.28	40
7.7	45	10.75	40
7.7	45	11.08	40
7.7	45	12.03	40
9.1	45	16	40
9.1	45	16.8	40
9.1	45	19.51	40
9.1	45	20.3	40
9.8	45	22.18	40
10.5	45	27.5	40
11.2	45	40.3	40
12.6	45	70.7	40
15.4	45	95	40
16.8	45	160	40
10.8	45	20.1	39
12	45	24.6	44
10.8	45	29.4	47
12	45	36.7	39
13.8	45	40.9	44
13.8	45	40.9	37
13.2	45	46	36
13.8	45	66.8	38
15	45	66.8	38
13.8	30	20.97	40
16.2	30	23	40
15	30	27.01	40

Figure 12.11 ESTIVATE sample data.

SUMMARY OUTPUT

Regression Statistics	
Multiple R	0.93925164
R Square	0.88219364
Adjusted R Square	0.8768388
Standard Error	4.47262482
Observations	70

ANOVA

	df	*SS*	*MS*	*F*	*Significance F*
Regression	3	9886.98911	3295.663	164.74713	1.37657E-30
Residual	66	1320.288605	20.00437		
Total	69	11207.27771			

	Coefficients	*Standard Error*	*t Stat*	*P-value*	*Lower 95%*	*Upper 95%*	*Lower 95.0%*	*Upper 95.0%*
Intercept	-5.26617978	1.619186737	-3.252361	0.001805	-8.49898924	-2.03337	-8.49898924	-2.03337032
X Variable 1	0.09222088	0.016434699	5.611352	4.296E-07	0.059407959	0.125034	0.05940796	0.125033806
X Variable 2	0.07562266	0.00946166	7.992537	2.73E-11	0.056731857	0.094513	0.05673186	0.09451347
X Variable 3	0.30920756	0.043751429	7.067371	1.229E-09	0.221855048	0.39656	0.22185505	0.396560077

Figure 12.12 Multiple regression output.

An example of the data is provided in Figure 12.11; although to obtain the complete source data and associated analysis, readers are advised to consult the original manuscript published in *Engineering, Construction and Architectural Management* (details are provided in the bibliography).

To conduct the analysis a similar procedure to that used to calculate bivariate regression is used. Click on the `Tools` drop down menu and select `Data Analysis` and then `Regression`, then click `OK`. In the `Input Y Range` insert 'A2:A71' and in the `Input X Range` insert 'B2:D71'. This will ensure that the variables in columns B, C and D are entered into the analysis. As per the previous example, ensure that the `Residuals` tick box is highlighted before clicking on `OK`. Data output from this analysis is presented in Figure 12.12.

The resulting equation (resulting from cells B17 to B20 'coefficients') is expressed as:

$$y = -5.2661 + (0.0922 \times \text{swing angle}) + (0.0756 \times \text{machine weight}) + (0.3092 \times \text{dig depth})$$

The r^2 value, 0.88, reveals that the model is a good predictor of machine cycle time and could be used to estimate machine cycle time.

Time series forecasting

Certain types of data are monitored and collected over regular time periods, for example, company profits, maintenance expenditure, fuel consumption and so forth. Such data are used to observe the changing patterns or trends throughout a series of given time periods and can be predicted to determine the future direction of the trend. There are various methods by which to predict a time series. The two that will be discussed here are the complementary techniques of autoregression and moving averages. These techniques can utilise quite complex multiple variables (multivariate analysis) or mono variables (bivariate analysis). For the purposes of simplicity, bivariate examples are given.

Autoregression

Autoregression can be used on time series data where the time series variable is related to itself. Therefore, the time series y_t (the dependent variable) is regressed on past values in the series itself, that is, $y_{t-1}, y_{t-2}, \ldots, y_{t-x}$ (the independent or predictor variable). The autoregression equation can be expressed mathematically as:

$$\hat{y} = b_0 + b_1 y_{t-1} + b_2 y_{t-2}$$

where

- b_0 is a constant.
- b_1 and b_2 are the autoregression coefficients.
- y_{t-1} and y_{t-2} are the previous two values of the time series trend.

This particular equation is more accurately called a second order autoregressive equation because it uses two lagged predictor variables at time y_{t-1} and y_{t-2}. However in mathematical terms any order can be used, hence the common convention refers to an nth order autoregressive equation, for example:

$$\hat{y} = b_0 + b_1 y_{t-1} + b_2 y_{t-2} + \cdots + b_n y_{t-n}$$

This technique is of particular use when forecasting short to medium range forecasts (normally one–four years) and when the series is not extremely volatile (i.e. where data values are erratic). For volatile series a more detailed analysis technique such as an Autoregressive Integrated Moving Average (ARIMA) model may be required or transformation of the data before applying the technique. Both of these are not considered any further in this work, but further and more detailed information can be found in Chatfield (1996).

Example 12.4

Records of the monthly downtime rate for a wheeled loader operating within the civil engineering industry have been recorded. Twenty-four monthly observations were recorded during the years 2000 and 2001 (refer to Figure 12.13).

To conduct the analysis click on the `Tools` drop down menu and select `Data Analysis` and then `Regression`, then click `OK`. In the `Input Y Range` insert ‘C5:C25’ and in the `Input X Range` insert ‘D5:F25’. Columns D to F (y_{t-1}, y_{t-2} and y_{t-3}) represent

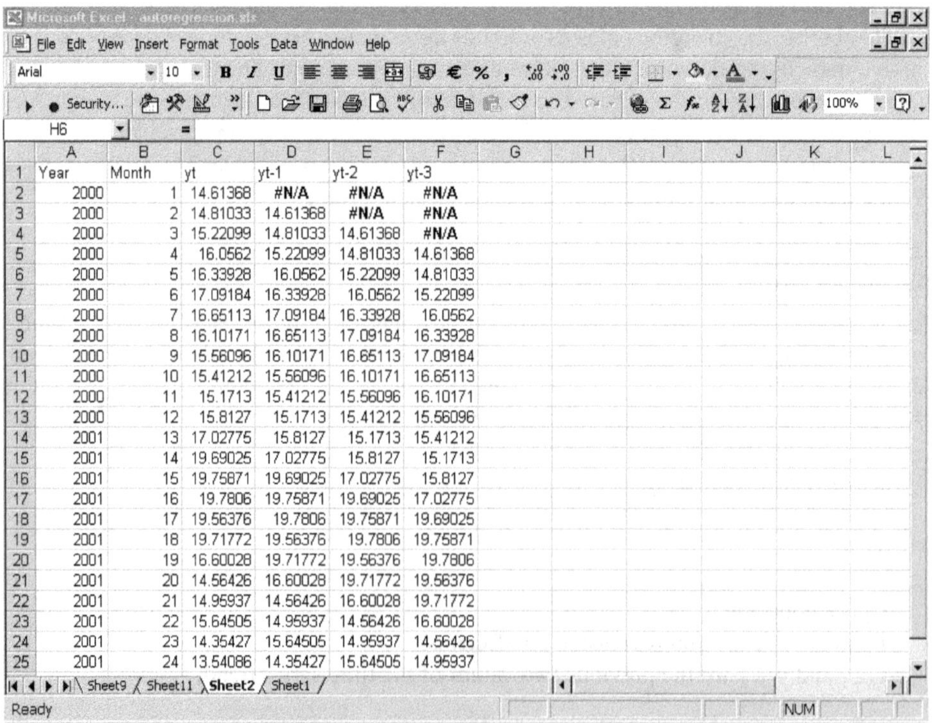

	A	B	C	D	E	F
1	Year	Month	yt	yt-1	yt-2	yt-3
2	2000	1	14.61368	#N/A	#N/A	#N/A
3	2000	2	14.81033	14.61368	#N/A	#N/A
4	2000	3	15.22099	14.81033	14.61368	#N/A
5	2000	4	16.0562	15.22099	14.81033	14.61368
6	2000	5	16.33928	16.0562	15.22099	14.81033
7	2000	6	17.09184	16.33928	16.0562	15.22099
8	2000	7	16.65113	17.09184	16.33928	16.0562
9	2000	8	16.10171	16.65113	17.09184	16.33928
10	2000	9	15.56096	16.10171	16.65113	17.09184
11	2000	10	15.41212	15.56096	16.10171	16.65113
12	2000	11	15.1713	15.41212	15.56096	16.10171
13	2000	12	15.8127	15.1713	15.41212	15.56096
14	2001	13	17.02775	15.8127	15.1713	15.41212
15	2001	14	19.69025	17.02775	15.8127	15.1713
16	2001	15	19.75871	19.69025	17.02775	15.8127
17	2001	16	19.7806	19.75871	19.69025	17.02775
18	2001	17	19.56376	19.7806	19.75871	19.69025
19	2001	18	19.71772	19.56376	19.7806	19.75871
20	2001	19	16.60028	19.71772	19.56376	19.7806
21	2001	20	14.56426	16.60028	19.71772	19.56376
22	2001	21	14.95937	14.56426	16.60028	19.71772
23	2001	22	15.64505	14.95937	14.56426	16.60028
24	2001	23	14.35427	15.64505	14.95937	14.56426
25	2001	24	13.54086	14.35427	15.64505	14.95937

Figure 12.13 Downtime sample data.

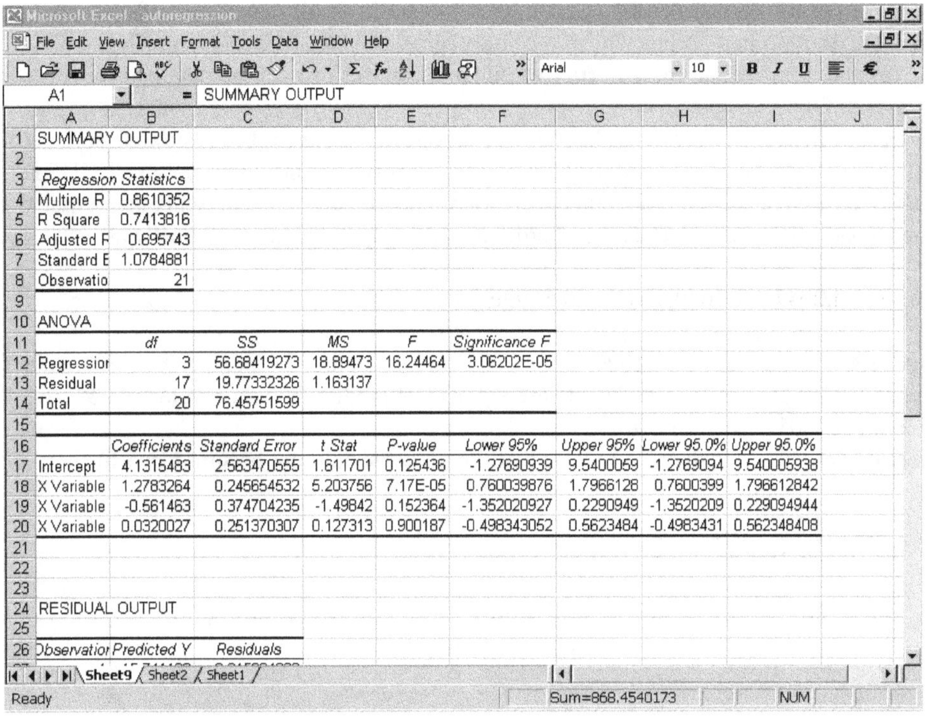

	A	B	C	D	E	F	G	H	I
1	SUMMARY OUTPUT								
2									
3	*Regression Statistics*								
4	Multiple R	0.8610352							
5	R Square	0.7413816							
6	Adjusted R	0.695743							
7	Standard E	1.0784881							
8	Observatio	21							
9									
10	ANOVA								
11		*df*	*SS*	*MS*	*F*	*Significance F*			
12	Regressior	3	56.68419273	18.89473	16.24464	3.06202E-05			
13	Residual	17	19.77332326	1.163137					
14	Total	20	76.45751599						
15									
16		*Coefficients*	*Standard Error*	*t Stat*	*P-value*	*Lower 95%*	*Upper 95%*	*Lower 95.0%*	*Upper 95.0%*
17	Intercept	4.1315483	2.563470555	1.611701	0.125436	-1.27690939	9.5400059	-1.2769094	9.540005938
18	X Variable	1.2783264	0.245654532	5.203756	7.17E-05	0.760039876	1.7966128	0.7600399	1.796612842
19	X Variable	-0.561463	0.374704235	-1.49842	0.152364	-1.352020927	0.2290949	-1.3520209	0.229094944
20	X Variable	0.0320027	0.251370307	0.127313	0.900187	-0.498343052	0.5623484	-0.4983431	0.562348408
21									
22									
23									
24	RESIDUAL OUTPUT								
25									
26	Observatior	*Predicted Y*	*Residuals*						

Figure 12.14 Autoregression output for the third-order model.

the series y_t at the value of one, two and three months previous respectively. As per the previous example, ensure that the `Residuals` tick box is highlighted so that model accuracy can be tested. Data output is presented in Figure 12.14.

The resulting equation is expressed as:

$$\hat{y} = 4.1315 + 1.2783y_{t-1} + -0.5614y_{t-2} + 0.0320y_{t-3}$$

where the origin is year 2000 (month 4) and Y units = 1 month. To test whether the equation is a significant predictor, the highest order coefficient must be tested; in this example, this is b_3. By convention the test is expressed as: H0: $b_3 = 0$ (b_3 is not a significant predictor) against H1: $b_3 \neq 0$ (b_3 is a significant predictor).

This is achieved by dividing the value of the coefficient b_3, from cell B20, by the standard deviation (standard error) of the coefficient (SDb_3) from cell C20. Hence, the test statistic (t) is as follows:

$$t = \frac{0.0320}{0.2513} = 0.1273$$

Using a 0.05 level of significance, the two-tailed t test with 17 degrees of freedom (Cell B13) has a critical value $t17$ of ± 2.1098 (taken from statistical tables).

Because $t = 0.01273 < 2.1098$, the upper-tailed critical value under the t distribution and because the p-value of 0.9001 (Cell E20) $> \alpha = 0.05$, we do not reject H0 and conclude that the third order parameter of the autoregressive model is not significant and can be deleted. Further and more definitive guidance on this test can be obtained from Levine *et al.* (2000).

Despite this result a second order model may prove to be more robust. The analysis is reconducted using the same methodology as before but this time including only cells 'D5:E25'; hence, creating a second order autoregression model. Figure 12.15 presents the output from the reconducted analysis. The second order model is expressed mathematically as:

$$\hat{y} = 4.2047 + 1.2640y_{t-1} + -0.5199y_{t-2}$$

where the origin is year 2000 (month 3) and Y units = 1 month. It is immediately apparent that the model coefficients are significant predictors with p-values of 5.77E − 06 (i.e. 0.000000577) and $0.020214 < \alpha = 0.05$.

However, the t statistic also reveals that the model should not be simplified further:

$$t = \frac{-0.5199}{0.2051} = -2.53$$

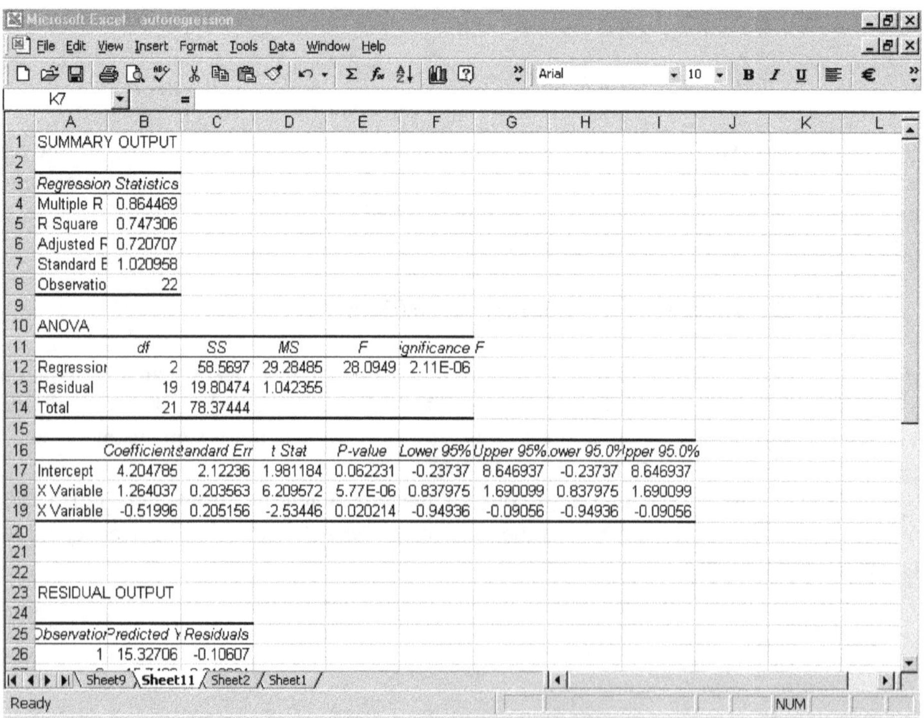

	A	B	C	D	E	F	G	H	I
1	SUMMARY OUTPUT								
2									
3	*Regression Statistics*								
4	Multiple R	0.864469							
5	R Square	0.747306							
6	Adjusted F	0.720707							
7	Standard E	1.020958							
8	Observatio	22							
9									
10	ANOVA								
11		*df*	*SS*	*MS*	*F*	*ignificance F*			
12	Regressior	2	58.5697	29.28485	28.0949	2.11E-06			
13	Residual	19	19.80474	1.042355					
14	Total	21	78.37444						
15									
16		*Coefficient*	*tandard Err*	*t Stat*	*P-value*	*Lower 95%*	*Upper 95%*	*ower 95.0%*	*pper 95.0%*
17	Intercept	4.204785	2.12236	1.981184	0.062231	-0.23737	8.646937	-0.23737	8.646937
18	X Variable	1.264037	0.203563	6.209572	5.77E-06	0.837975	1.690099	0.837975	1.690099
19	X Variable	-0.51996	0.205156	-2.53446	0.020214	-0.94936	-0.09056	-0.94936	-0.09056
20									
21									
22									
23	RESIDUAL OUTPUT								
24									
25	*Observatio*	*Predicted Y*	*Residuals*						
26	1	15.32706	-0.10607						

Figure 12.15 Autoregression output for the second-order model.

Making the prediction

The autoregression modelling process has determined that the second-order model is the most robust model given the available data. Using the estimates $b_0 =$, $b_1 =$, and $b_2 =$ as well as the two most recent data values, months 23 (14.35427) and 24 (13.54086), the forecasts for month 25, then months 26 and 27 are:

$$\text{Month 25: } \hat{y}_{25} = 4.2047 + 1.2640(13.54086) + -0.5199(14.35427) = 13.85756$$
$$\text{Month 26: } \hat{y}_{26} = 4.2047 + 1.2640(13.85756) + -0.5199(13.54086) = 14.68076$$
$$\text{Month 27: } \hat{y}_{27} = 4.2047 + 1.2640(14.68076) + -0.5199(13.85756) = 15.55664$$

The final interpretation of the results is largely subjective and dependent upon the knowledge and experience of the analyst. However, for this example it is reasonable to assume that the downtime rate of a machine is cyclical in that it increases and then decreases as major overhaul works are conducted.

Moving averages (or smoothing a time series)

Moving averages is a simple technique that can smooth the peaks and troughs of a turbulent trend and thus allow modelling of the 'smoothed' data to commence. The moving averages equation is expressed as:

$$\text{MA}_n = \frac{Y_1 + Y_2 + Y_3 + Y_n}{n}$$

where

- MA_n is the moving average (n) model.
- Y is the independent variable observed at points one to n.
- n is the number of observations taken.

Therefore, a three-period moving average would be represented as:

$$\text{MA}_{(3)} = \frac{Y_1 + Y_2 + Y_3}{3}$$

Example 12.5

Figure 12.16 provides details of a civil engineering firm's profits over a period of 24 years (1977–2000). The finance department needs to make a prediction of future company profits using bivariate regression analysis but has revealed that the trend in profits is erratic and thus apparently unpredictable (refer to Figure 12.17).

In an attempt to smooth this trend, the MA_3 `(3p)`, MA_4 `(4p)` and MA_5 `(5p)` year period trends are plotted and examined.

For the calculation of the 3p trend, enter the formula `=sum(B2:B4)/3` in cell C4, and replicate for all Column C cells. Similarly, for the 4p and 5p trends enter the formula `=sum(B2:B5)/4` in cell D5 and `=sum(B2:B6)/5` in cell E6 respectively and replicate for other cells in columns D and E. The resulting trends are plotted in Figure 12.18.

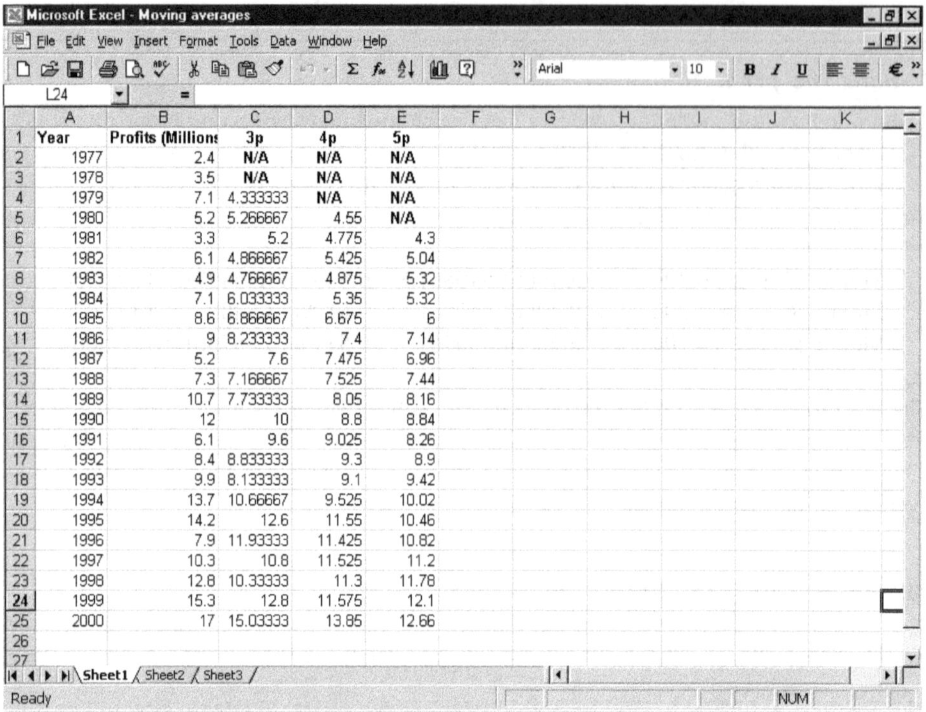

	A	B	C	D	E
1	Year	Profits (Millions	3p	4p	5p
2	1977	2.4	N/A	N/A	N/A
3	1978	3.5	N/A	N/A	N/A
4	1979	7.1	4.333333	N/A	N/A
5	1980	5.2	5.266667	4.55	N/A
6	1981	3.3	5.2	4.775	4.3
7	1982	6.1	4.866667	5.425	5.04
8	1983	4.9	4.766667	4.875	5.32
9	1984	7.1	6.033333	5.35	5.32
10	1985	8.6	6.866667	6.675	6
11	1986	9	8.233333	7.4	7.14
12	1987	5.2	7.6	7.475	6.96
13	1988	7.3	7.166667	7.525	7.44
14	1989	10.7	7.733333	8.05	8.16
15	1990	12	10	8.8	8.84
16	1991	6.1	9.6	9.025	8.26
17	1992	8.4	8.833333	9.3	8.9
18	1993	9.9	8.133333	9.1	9.42
19	1994	13.7	10.66667	9.525	10.02
20	1995	14.2	12.6	11.55	10.46
21	1996	7.9	11.93333	11.425	10.82
22	1997	10.3	10.8	11.525	11.2
23	1998	12.8	10.33333	11.3	11.78
24	1999	15.3	12.8	11.575	12.1
25	2000	17	15.03333	13.85	12.66

Figure 12.16 Company profits.

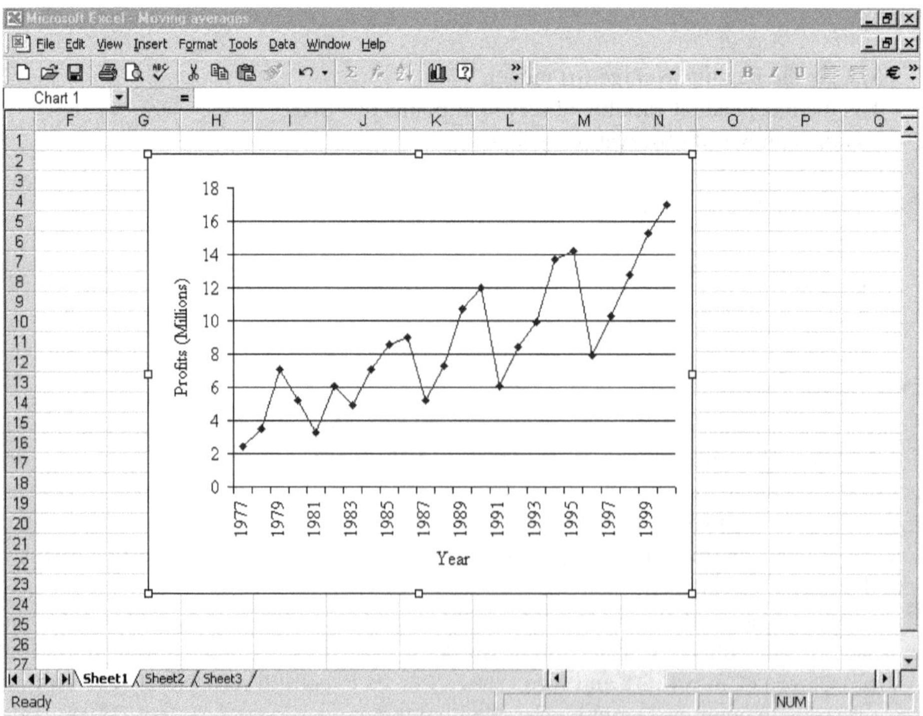

Figure 12.17 The trend in company profitability.

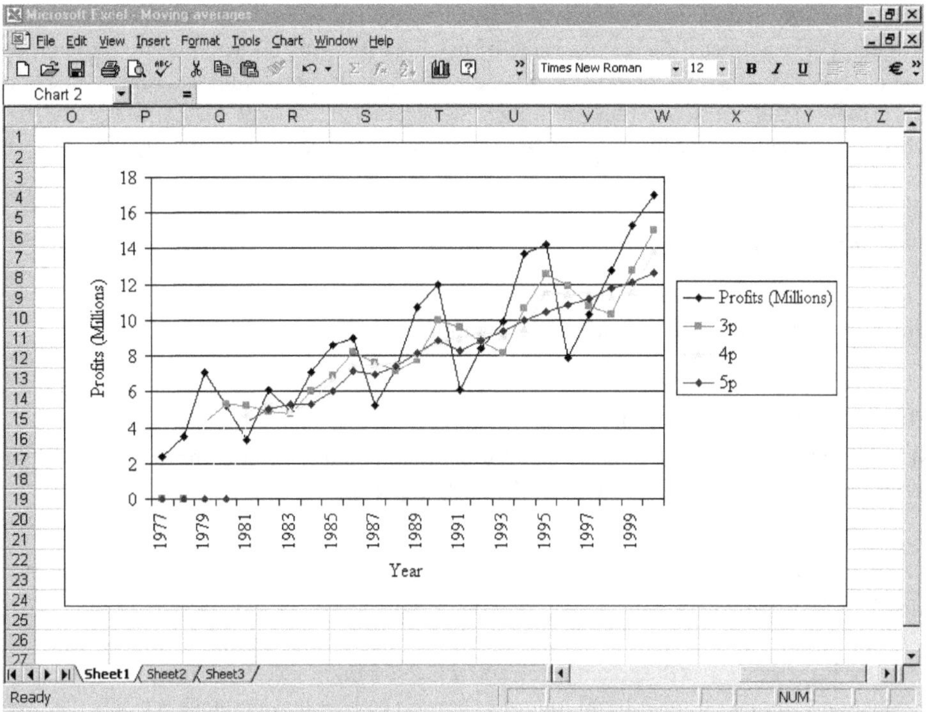

Figure 12.18 MA_3 to MA_5 smoothing trends.

It is clear that with each additional smoothing from MA_3 to MA_5, the trend slowly becomes more linear. Indeed, the MA_5 is virtually linear and on calculation has an r value of 0.81. This high level of association between actual profits and the MA_5 data proves that profits (the dependent variable) could be successfully modelled using the MA_5 data (the independent variable) in bivariate regression analysis.

Performance analysis and judgement

Modelling data using the aforementioned techniques reveals that from a statistical perspective, a reasonable prediction of future trends can be made. However, prior to utilising such a model in the development of the operational plan, the performance of the model must be analysed and a judgement made as to how accurate the predictions are. This will engender confidence that use of the model developed will produce a reliable and robust forecast. An accepted method of performance analysis is to undertake an analysis of the residuals; where a residual value is the difference between the actual and forecasted value. This can be achieved using various tests amongst which are the Mean Absolute Deviation (MAD) and the Mean Absolute Percentage Error (MAPE). Note that some other tests look for compliance with statistical prerequisite requirements (e.g. normality in parametric models) these are beyond the scope of this book.

Mean absolute deviation

Mean absolute deviation represents the absolute deviation from the prediction for each residual. For example, if the actual value of machine sales is 200 machines and the prediction is 150 machines then the error is 50. Conversely, if the prediction is 250 machines then the error is still 50. Thus, plus or minus signs are not used in MAD, hence the term 'absolute' deviation. The formula for MAD can be expressed mathematically as:

$$\text{MAD} = \frac{\sum |e|}{n}$$

where $|e|$ is the absolute error and n is the sample size.

Mean absolute percentage error

MAPE represents the relative error of each prediction and is most commonly used in time series analysis and regression studies. MAPE is expressed mathematically as:

$$\text{MAPE} = \frac{\sum \left|\frac{e}{y}\right|}{n}$$

where: e is the error, y is the actual value and n is the sample size. For example, if a machine production forecast estimated 200 tonnes of topsoil to be excavated and yet only 190 were excavated then the absolute percentage error would be:

$$\left|\frac{190 - 200}{200}\right| = 0.05$$

Therefore the error for this prediction is 5 per cent of the actual value. The MAPE is the sum of the absolute errors over the range of predictions made. With both MAPE and MAD there is nothing precluding the use of median values as opposed to mean. Indeed, for skewed distributions, the median value is the preferred measure of central tendency.

Judgement

Having vigorously tested the scientific validity of the results, managers will be confronted with the decision regards whether, or not, to accept the model and its predictions. Under this situation, the manager's qualitative judgements provide a sufficient base upon which to accept or reject the model. Judgement value (i.e. the extent to which the individual judgement can be trusted) largely depends upon the extent of a manager's experience and success. Often, models and performance analysis are judged on the consensus of opinion of a group of experts from various operational areas within the company (e.g. purchasing, finance, contracts, plant management and so forth) so as to ensure a diverse and in-depth knowledge base. Any modelling estimates made are then adjusted for any anticipated changes (events) that may impact upon model accuracy. The main strengths of model validation by judgement are that it is comparatively inexpensive to conduct, it communicates predictions throughout the company and commits management to them and it combines the very best features of intuition and scientific forecasting.

Gantt charts

Having obtained accurate models of the various types of data collected (maintenance costs, machine production and so forth) the organisation's operational plan for plant management can be devised and implemented. To ensure that this plan is adhered to and monitored adequately requires the use of planning techniques. These may include: programming networks such as Performance Evaluation and Review Techniques (PERT), Line of Balance and so on. However, the most widely used technique within plant management practice is the Gantt chart (named after the American inventor H.L. Gantt) because of its simple construction and ease of interpretation.

Gantt charts consist of a series of bars located on a 'time line' chart. Each bar represents an activity (e.g. fixed-time-to maintenance work), a duration (hours, days or weeks) to complete that activity, a start date and an expected completion date. Progress of the activity may then be monitored to ensure that it is completed on schedule. A logical relationship is established between activities by the use of links, normally represented by arrows. There are four types of activity link. These are:

1. *Start, finish* (SF) where an activity starts when its predecessor finishes.
2. *Finish, finish* (FF) where two activities must be completed at the same time.
3. *Start, start* (SS) where two activities commence at the same time.
4. *Finish, start* (FS) where an activity must be completed when its predecessor commences (seldom used).

These links indicate precise interrelationships between activities. It is important that activities are plotted in a logical and practical sequence so that each activity follows in the correct order. It should also be noted that it only takes one activity to be completed late for the whole works to be disrupted and delayed. Figure 12.19 presents a simple Gantt chart, produced using Microsoft Project for the overhaul of a hydraulic excavator.

Ten 'main' activities (or task names) are shown, each with a given duration, start date and finish date. Using a combination of knowledge acquired from previous maintenance works conducted, statistics and intuition the plant manager must first estimate the duration of each activity. At this stage, care should be taken to ensure that some time allowance is made for unexpected problems that may occur such as mechanic absenteeism, shortage of parts and consumables or identification of additional works that may be required. *It is far better to set a reasonable duration and achieve it than set an ambitious duration and fail to achieve it*! Once these durations have been reliably estimated, the plant manager must then:

1. Establish the order and manner in which the works are to be undertaken.
2. Identify activities that must be completed within a given time and are constrained by other activities (in strict interlocking sequence). This sequence of activities is aptly called the critical path and must be completed on time in order for the whole project to be completed on time. Any delay in completing critical path activities will result in project delay.
3. Determine activities that have a degree of latitude in time (i.e. they can be completed within a given time frame). This additional time allowed to complete an activity is commonly known as the float.

Logical links between the activities can then be made (refer to Figure 12.20) and a reliable programme of works carefully prepared.

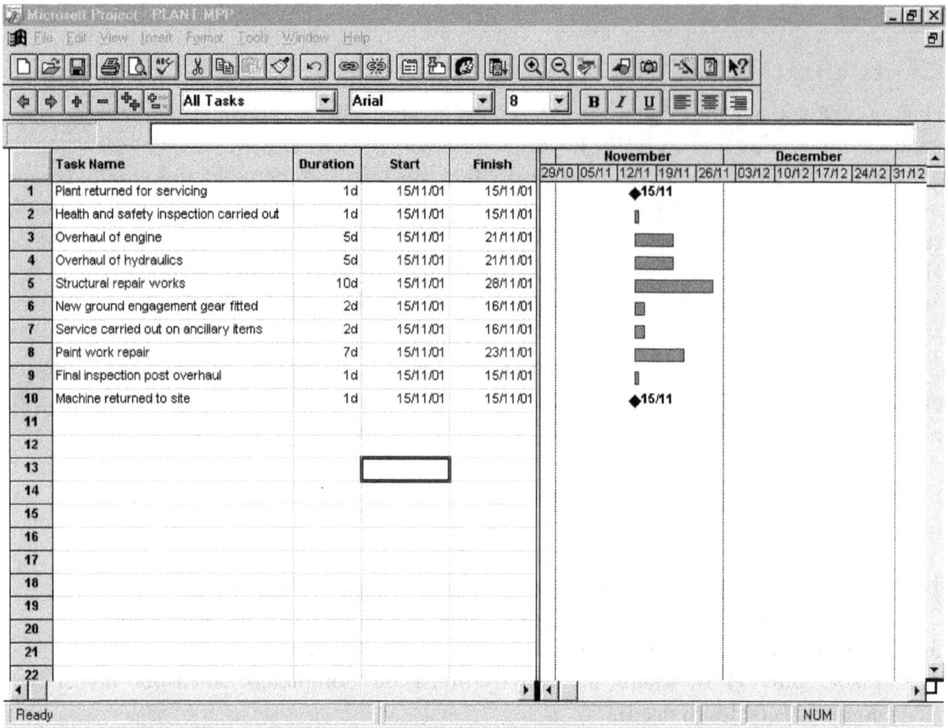

	Task Name	Duration	Start	Finish
1	Plant returned for servicing	1d	15/11/01	15/11/01
2	Health and safety inspection carried out	1d	15/11/01	15/11/01
3	Overhaul of engine	5d	15/11/01	21/11/01
4	Overhaul of hydraulics	5d	15/11/01	21/11/01
5	Structural repair works	10d	15/11/01	28/11/01
6	New ground engagement gear fitted	2d	15/11/01	16/11/01
7	Service carried out on ancillary items	2d	15/11/01	16/11/01
8	Paint work repair	7d	15/11/01	23/11/01
9	Final inspection post overhaul	1d	15/11/01	15/11/01
10	Machine returned to site	1d	15/11/01	15/11/01

Figure 12.19 Gantt chart.

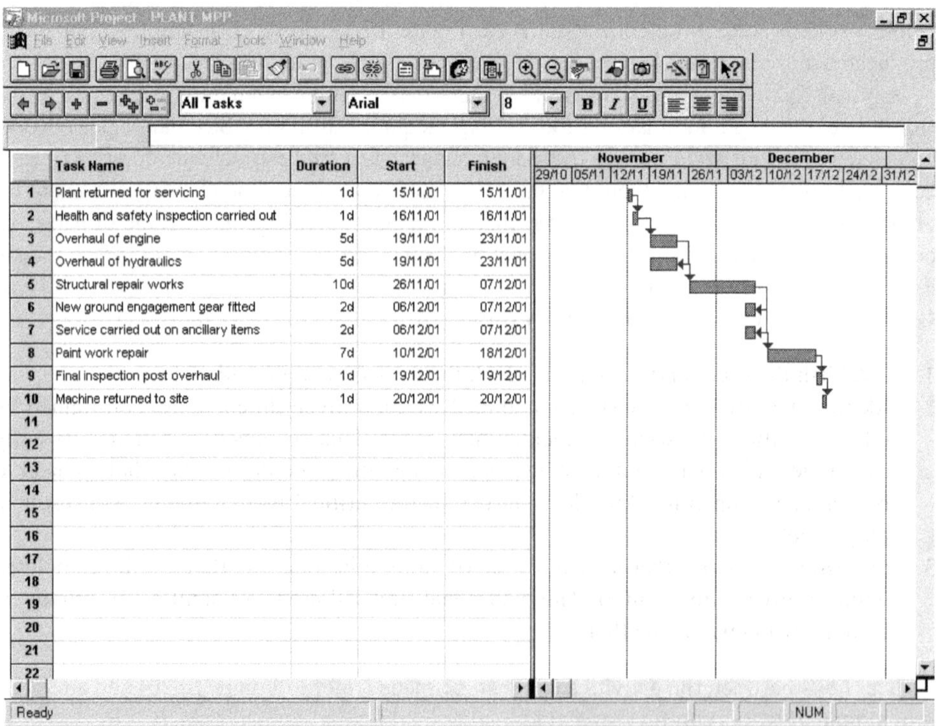

	Task Name	Duration	Start	Finish
1	Plant returned for servicing	1d	15/11/01	15/11/01
2	Health and safety inspection carried out	1d	16/11/01	16/11/01
3	Overhaul of engine	5d	19/11/01	23/11/01
4	Overhaul of hydraulics	5d	19/11/01	23/11/01
5	Structural repair works	10d	26/11/01	07/12/01
6	New ground engagement gear fitted	2d	06/12/01	07/12/01
7	Service carried out on ancillary items	2d	06/12/01	07/12/01
8	Paint work repair	7d	10/12/01	18/12/01
9	Final inspection post overhaul	1d	19/12/01	19/12/01
10	Machine returned to site	1d	20/12/01	20/12/01

Figure 12.20 Linked Gantt chart.

In the example used, the plant item must be returned for servicing (activity 1) before the health and safety assessment is conducted (activity 2); hence the FS link between these activities. Having completed the assessment, overhaul of the engine and hydraulics (activities 3 and 4) can commence, and given sufficient fitter support, these activities can commence and complete at the same time; hence the FF link. For this example, only main activity headings have been included and these could be broken down into further detail if necessary (that is, sub-activity headings). The engine and hydraulic overhaul activities represent typical examples of such since oil sampling, filter replacement, oil replacement and so forth could have been included under these generic section headings.

Advantages of the Gantt chart

The main advantages of the Gantt chart are the:

1 Inherent simplicity to produce and interpret.
2 Ease of recording progress and rescheduling activities.
3 Flexibility in that 'short term' more intricate programmes can be extracted from the main programme.
4 Depth of information and logical structure, enabling for example, data for ordering materials (and other resources) to be easily identified and managed.

These advantages are compelling and have secured the success of the Gantt chart in modern plant management science.

Operational planning quality

At the outset of this chapter, information was highlighted as being of the utmost importance to the development of an effective 'plan'. Once the plan has been implemented, information is again required to judge its effectiveness, that is, if it has achieved the desired objectives and has been-accurate. Elements upon which the plan should be assessed include:

1 Its user-friendliness for both managers and employees.
2 The attention to detail covered and whether all eventualities were considered.
3 The relative success of the plan as measured by its ability to meet desired objectives, that is, did it work.
4 The inherent flexibility and stability of the plan. For example, was the plan expressed and defined clearly (stability) and did the plan allow flexibility in the light of changing circumstances? This may include incorporating contingency plans to supplement existing plans.

An effective monitoring mechanism is required to measure the relative success (i.e. when compared to previous plans) of each of the above elements. This is not a difficult task, once relevant performance measures have been established, and is primarily a clerical activity. Nevertheless, a person or department (entity) within the organisation should have a clear, delegated responsibility for quality. Further, it is important that the entity is connected to the production and/or operations department. This is because the 'watch dog' nature of quality control can cause friction within the organisation and possibly lead to morale and cooperation problems. For example, change in operations management can be perceived by line managers as interference and thus resented. In turn, line managers may positively resist change 'improvements' and actively seek to undermine them.

The performance method of measurement used is dependent upon the nature and purpose of the plan, for example, is the plan concerned with finance, production, human or material resources? Regular dissemination of the plan's performance ensures natural stability and continuity of the plan. This can be easily achieved using the open architecture feature found on many commercially available software packages, e-mail, the internet or a mixture of all three.

Chapter 13

International operations

Introduction

Well established organisations regularly undertake work and commercial developments overseas, particularly in developing countries and increasingly in nearby regional markets, for example, the European Union (EU), SE Asia, North America, etc. Projects which have engaged contractors include housing, roads, power stations, water provision, treatment and disposal, mining, pipelines, agricultural development and other infrastructure activities. The latter have included contracts for electro-mechanical complexes, steelworks and other industrial facilities, especially for the multinational manufacturers. Most have required changes in management and procurement practices in dealing with operations remote from or foreign to the home base, not least relating to the logistics of equipment supply and usage.

Logistical problems

A remote region usually implies locations in uninhabited territory and sparsely populated areas lacking basic infrastructure and utilities, such as good transport networks and communications. However, unfamiliar conditions can occur almost next door, resulting from different languages, and political, economic or historical developments. Nevertheless, in general, operations will mainly involve changed geography, climate, and legal and/or unaccustomed social practices. The 'comfortable' home arrangements of immediate deliveries, spares 'off the shelf', ready advice and abundant experienced staff, etc., are unlikely to be available.

Many of the logistical problems to be faced in supplying and operating equipment in remote regions are common to well-developed societies but usually require considerably more management effort. Typically, wrong decisions and choices do occur and, as a consequence, already overstretched resources become exacerbated, the situation being further compounded by scheduling, ordering, despatch and, especially, delivery delays. Indeed, systems designed for a developed home market often break down when exposed to the vagaries of the different communications and infrastructure of a foreign environment. Other difficulties commonly arise through the need to raise or receive funds in a foreign currency, which sometimes may not be readily acceptable to the Foreign Exchange market. Also, import controls and customs regulations can be complicated and very restrictive, particularly regarding imported spare parts and especially where the host country claims to have its own manufacturing capability.

Obtaining the right personnel to both manage and operate equipment may prove difficult. Indigenous staff often require considerable training, both in the skills for the job itself and in communicating with the home base and systems of the company, including the need to converse in a foreign language. Expatriates often predominate at supervisory levels and, unless highly paid, thoroughly selected and prepared for the foreign environment, are unlikely to perform satisfactorily. Furthermore, having to cope with local customs and practices commonly borders on corruption and can become very distorted unless entrusted to very reliable staff.

More specifically, unfamiliar communication facilities, such as telephone systems, postal services and even radio transmission, can lead to frustration and errors. Regarding equipment itself, inexperienced staff can inadvertently cause extra maintenance, increased breakdowns and failures, etc., further compounded by damages to equipment in transit, either through bad roads and transport or even lack of adequate inspection of imports at source. Also, extra quantities, especially of spare parts, may need to be ordered, to avoid stock shortages. Furthermore, overseas work frequently exposes inefficient manufacturers and suppliers, with many unprepared for dealing with unfamiliar logistics and markets.

Equipment requirements

Support items

The types of equipment needed to execute an overseas project will clearly depend upon the nature of the work. However, support facilities can be more specifically mentioned and would typically include:

1. Concrete batching and mixing equipment.
2. Transport vehicles, such as trucks, four-wheel Land Rovers, utility vans and wagons.
3. Welding and fabrication shops, equipped with welding and burning gear, workshop equipment, grinders, etc.
4. Maintenance workshops, for plant and vehicles, together with spare parts stores.
5. Power generators, including a standby water storage tank, fuel dumps, etc. A water treatment capability might also be necessary.
6. Waste disposal facilities for sewage, for example, septic tanks and incinerators. Indeed, on large long-term sites proper sewage treatment equipment may be worth installing.
7. Furnished site offices, canteens, stores and compound, plant yard. Many sites will also need to provide furnished accommodation for some of the work force, which might have to be quite elaborate for expatriates, including recreational facilities.

Equipment in general

Wherever possible, the equipment acquired or designed for an overseas contract should be versatile, with the capacity to change one duty for another, for example from tractor to carrier. Standardisation can assist here in facilitating switching and cannibalisation. Sensitive equipment generally needs protection from the heat or cold, dust, humidity, wind and other climatic conditions, with instruments requiring special attention, particularly during transport. Temperature-sensitive materials should be kept under controlled conditions in cold storage or air-conditioned rooms.

Above all, regular maintenance is essential, even to the extent of regularly turning on mechanical and electrical components to keep moving parts free. A plentiful supply of stocks of consumables, such as steel plate, structural steel pieces, pipework, hosing and other fittings, is clearly necessary besides normal plant spares, and, above all, a well-protected and well-policed stores is essential. Evidence indicates that about 10–15 per cent of equipment value in spares is usually required as stocks where the local market is relatively economically unsophisticated/underdeveloped. Furthermore stock requirements may fluctuate markedly over a contract, and therefore a good-size initial supply is prudent until usage patterns have been ascertained. Local supplies may be available but could be inferior if not manufactured in one of the major industrial countries. This whole question impacts on maintenance, since, if the major manufacturers of equipment have no service backup in the country, either local fitters, electricians, mechanics and technicians have to be trained, or imported expatriates have to be employed at considerable expense. In either case, decisions regarding provision of appropriate spares and workshops must be carefully thought out.

Supply and delivery

While many of the above points affect the selection of equipment and the arrangements needed to effectively manage and operate, other factors such as supply, packaging, transport, agents, customs, finance, insurance, and so on, are equally important.

Pre-shipment inspection

Equipment is commonly specified to certain standards – for example, ISO, EN, BS, DIN, etc. and for the EU, CE marking of equipment plus much other certification – with local clients insisting on pre-shipment inspections being charged to the supplier, to ensure that the conditions are met before items leave the source country. The procedures requested may involve simply counting of items and dimensional checks, or more thorough inspections, including testing of materials, components and complete items of equipment, in accordance with the particular specification standard. Other aspects could include monitoring the export price against like items sold on the home market.

Transport to site

Transport over long distances through unfamiliar territory and other countries en route clearly needs special attention and planning. Wherever transit periods exceed several months and different means of carriage, strong packaging and preferably containerisation is necessary. The choice of transport method itself depends upon several factors – for example, roads, while generally economically competitive, usually involve long journey hours and can be fraught with problems, such as poor road surfaces, lack of signposting, pilferage, banditry, and unreliable drivers and/or haulage firms. Also, when at journey's end, special temporary roads may need to be constructed, especially for the very remote construction site.

Alternatively, shipment over long sea routes, while generally cheap, is very slow but is commonly preferred for large plant items, to avoid disassembly demanded by container sizes. However, even with shipping, containerisation needs link-ups with trucks for ease of handling. A further consideration is the quality of facilities at the inloading port. Increasingly, rail transport is becoming competitive over long distances especially where national rail systems are well organised – for example, Trans-Siberia, Trans-Canada and systems in parts of Africa and Europe.

Air freight is usually the quickest means of transport but also the most expensive. Heavy or bulky items are usually unsuitable, with large units often needing disassembling. Occasionally very specialist transport can be considered, such as helicopters, and even more exotic means such as snowmobiles, pack animals and so on. Finally, the courier should not be overlooked for small items facilitating considerable shortening of customs delays.

Documentation and bureaucracy

The bureaucracy of the importing country can be complex and certainly needs to be understood. For example, in addition to the specification and quality inspections mentioned above, other barriers such as tendering and trading procedures, legislation relating to labour employment and the environment, road transport regulations and restrictions regarding particular imports can cause problems. The advice of a local agent would certainly be helpful for the newcomer to the market, and every effort should be made to pre-empt difficulties.

Useful advice can be obtained from Government and specific industry export organisations, such as the Overseas Trade Board, export intelligence services and the Export Group for the Constructional Industries. As far as documentation is concerned, a local representative in the destination country is essential, not least to deal with language interpretation. Indeed, a local agent of this sort may be necessary in each country through which the goods have to pass. Good practice may also require the employment of a shipping agent to expedite and co-ordinate orders and deliveries, and to be responsible for packing lists, bills of lading, shipping, customs forms, currency payments and so on. A good shipping agent will normally encourage marking of items with part numbers and the provision of detailed descriptions on the documentation to aid both checking and inspection, but also to enable follow-up enquiries when matters become aggravated through pilferage and/or transport deterioration.

Finance and payment

Equipment purchased direct for overseas contracts usually requires the organising of import licences and, of course, the mechanisms of payment. Furthermore, the plant may have to be fully depreciated during the contract and simply left in the host country, although in some countries small markets for the sale of second-hand equipment exist. Notably import priorities in developing countries are commonly solved by rationing through a waiting list, sometimes tempered by importance of use of the currency, for example equipment over consumer goods. Some transactions in local currency may be available via the foreign exchanges, but for very poor or ill-managed economies, few opportunities exist to exchange 'hard' (i.e. Euros, dollars, pounds, etc.) for 'soft' currency, since such currency can really only be used to make purchases in the local economy. Some international bartering may be possible, and today organisations specialising in exchanging commodities for consumer and capital goods do exist.

Much difficulty can be avoided when projects are funded by aid agencies such as the World Bank, United Nations, Bi-Lateral aid agencies, etc.; here a soft loan is usually involved – that is, a loan in a hard currency with low interest rate or easy payment terms, with the stipulation being sourcing of equipment from a particular country. Finally, import duties should not be overlooked – national agreements may have been negotiated through the WTO (World Trade Organisation), regional economic national groupings etc. Also, some countries waive the import duty if the equipment is to be re-exported at the end of its duties.

Conclusion

Clearly, when embarking on any overseas contracts, obtaining knowledge about the likely conditions to be faced is vital. Advice is available in the home country through government export agencies, trade associations, research & development organisations, shipping agents, private consultancy firms and so on. However, reconnoitring the local environment is paramount and in addition, for many countries, local agents, distributors, contacts, etc., are absolutely essential. In some cases forming partnerships or joint companies may be justified in order to secure knowledge of: (1) legal, financial and banking procedures; (2) import legislation, quality specifications and standards; (3) customs, shipping, delivery and distribution procedure and (4) maintenance expertise and labour management.

Part IV

Financial and budgetary control

Chapter 14

Budgetary control and costing

Introduction

A budget acts as a standard of measure against which actual performance may be compared. The Institute of Cost and Management Accountants defines budgetary control as 'the establishment of budgets, relating the responsibility of executives to the requirements of a policy and the continuous comparison of actual with budgeted results, either to secure by individual action the objective of that policy, or to provide a basis for its revision'. Budgetary control therefore involves:

- setting targets;
- monitoring progress;
- taking corrective action when necessary.

Within a system of budgetary control, budgets are established to bring together the financial requirements for the component parts of the enterprise over the forthcoming 12-months period. They are expressed as forward estimates of costs and revenues, usually derived from records of past performance adjusted for future expectations.

Preparation of budgets

The budgetary system comprises many individual budgets which are ultimately integrated into a master budget. The master budget (Table 14.1) is similar in format to a profit and loss account but, unlike the latter, is based on forward estimates of costs and revenues and is therefore only a forecast of the anticipated profit to be earned. From such an estimate other factors related to future expectations may be projected, such as the rate of return on capital employed, dividends to shareholders, capital to be retained in the business for reinvestment in assets and similar items related to profitability.

At the start of any attempt to prepare budgets for business activity during the coming year, it is first necessary to prepare a budget for the investments to be made in plant and equipment, since it is through these assets that revenues and costs are mostly generated. Subsequently, a sales and operating budget may be synthesised and the cash flow requirements determined. By a gradual process of reiteration budgets are modified to keep within the constraints on financial resources available, culminating finally in the master budget, as shown in Figure 14.1.

Table 14.1 Master budget

Code	*Item*	*Annual (£)*	*Weekly (£)*
Budgeted sales (R)	Plant hire	400,000	8,000
Total budgeted sales		400,000	8,000
Budgeted costs			
(W)	Workshop dept costs	100,000	2,000
(T)	Transport dept costs	55,000	1,100
(H)	Hire dept costs	20,000	400
(C)	Costing dept costs	25,000	500
(A)	Administration costs	15,000	300
(B)	Buying dept costs	25,000	500
(F)	Accounts dept costs	20,000	400
Total budgeted costs		260,000	5,200
Budgeted trading profit (sales minus costs)		140,000	2,800
Budgeted depreciation on plant and premises		100,000	2,000
Budgeted net profit before interest and tax		40,000	800

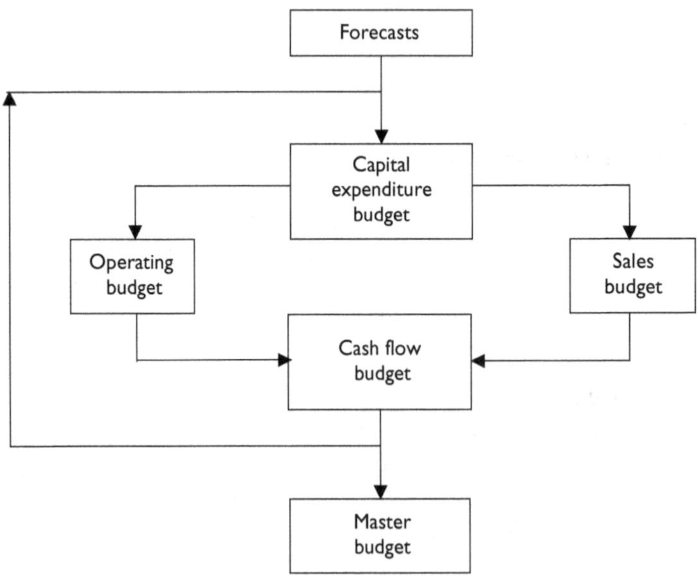

Figure 14.1 The master budget.

Types of budget

Capital investment budget

The capital investment budget is determined by the availability of capital and is thus a schedule of loan capital, equity, leases and retained profits. These sources of funds are treated more fully in Chapter 16, and the preparation of such a budget is covered in detail in Chapters 5 and 6.

Cash flow budget

The cash flow budget shows the short-term cash available period by period and is broadly determined by subtracting costs from revenues after taking into account payment delays. The preparation of a cash flow budget is described in detail in Chapter 15.

Sales budget

The sales budget for a rental or hire business is simply the forecast of revenues from equipment hire. This will be made up of the expected income from the hire of individual items, which may fluctuate on a seasonable basis. Consequently, the budget should be prepared showing the anticipated annual and weekly incomes.

Operating budget

The operating budget is prepared from estimated costs of the planned requirements for items such as:

- materials, for example, fuel, lubricants and spare parts;
- staff and labour;
- equipment such as small tools;
- depreciation of workshop equipment;
- business facilities, rent, rates, electricity, etc.

The difference in value between the sales and operating budgets is the anticipated profit before deduction of depreciation of the assets, such as equipment and buildings, etc., for the year ahead.

The operating budget, when appropriate, is subsequently subdivided into separate functions (see Figure 14.2). These may include budgets for departments such as transport, workshop and administration. In larger concerns the latter could be broken down further into hiring, sales, costing, accounts and purchasing. By providing each department with a separate financial budget, a target is available against which subsequent performance may be monitored. Examples of the form of the annual budgets for the workshop, transport and administration functions are shown in Tables 14.2–14.4. It can be seen that the budgets consist of cost forecasts of the requirements for materials, labour and expenses. A coding system is used to allocate the resources to particular departments or functions concerned, and like items are collected under the same alphanumeric code. It is usual to present both

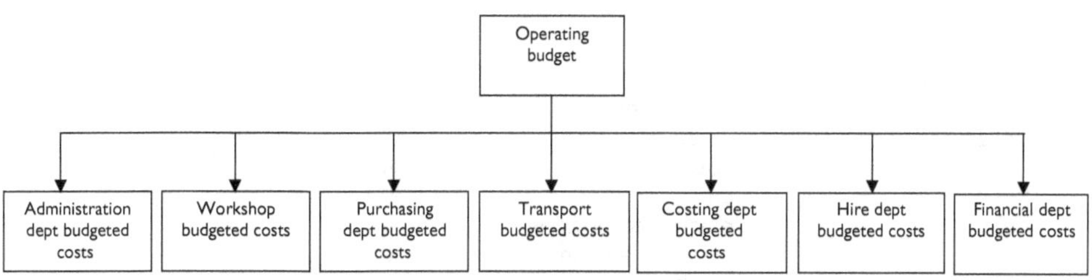

Figure 14.2 The operating budget.

Table 14.2 Workshop budget

Code (W)	*Item*	*Annual (£)*	*Weekly (£)*
Direct repair material and stock of spares			
10.0	Crawler cranes	6,000	120
20.0	Tower cranes	8,000	160
30.0	Trucks	4,000	80
40.0	Bulldozers	5,000	100
50.0	Loaders	7,000	140
60.0	Excavators	10,000	200
Cost		40,000	800
Direct labour costs			
10.1	Crawler cranes	7,000	140
20.1	Tower cranes	7,000	140
30.1	Trucks	5,000	100
40.1	Bulldozers	5,000	100
50.1	Loaders	8,000	160
60.1	Excavators	8,000	260
Cost		40,000	800
Indirect costs			
90.1	Storekeeper	6,000	120
90.2	Staff salaries	8,000	160
90.3	Rent	3,000	60
90.4	Rates	2,000	40
90.5	Electricity	500	10
90.6	Depreciation of tools and equipment	500	10
Cost		20,000	400
Total cost		100,000	2,000

Table 14.3 Transport budget

Code (T)	*Item*	*Annual (£)*	*Weekly (£)*
Direct material costs			
100.0	Vans	4,000	80
110.0	8-ton lorries	5,000	100
120.0	16-ton lorries	5,000	100
130.0	32-ton lorries	3,000	60
140.0	Cars	3,000	60
Cost		20,000	400
Direct labour costs			
100.1	Vans	3,000	60
110.1	8-ton lorries	6,000	120
120.1	16-ton lorries	5,000	100
130.1	32-ton lorries	4,000	80
140.1	Cars	2,000	40
Cost		20,000	400
Indirect costs			
190.1	Staff salaries	11,000	200
190.2	Electricity	500	10
190.3	Rates	1,000	20
190.4	Rent	1,000	20
190.5	Vehicle depreciation	2,000	40
190.6	Tools and equipment	100	2
190.7	Telephone, etc.	400	8
Cost		15,000	300
Total cost		55,000	1,100

Table 14.4 Administration budget

Code (A)	*Item*	*Annual (£)*	*Weekly (£)*
Direct employment costs			
300.0	Staff salaries	9,500	190
Direct material costs			
310.0	Stationery	200	4
Direct expenses			
320.0	Photocopying	100	2
Indirect costs			
330.1	Telephone	1,000	20
330.2	Postage	700	14
330.3	Electricity	1,000	20
340.1	Rates	500	10
340.2	Rent	1,000	20
340.3	Office equipment	400	8
340.4	Insurances	600	12
Total cost		15,000	300

an annual budget and a weekly budget, as short-term/seasonal fluctuations may be the more usual pattern of expected performance.

Classification of costs

For the preparation and monitoring of budgets, costs are collected and classified into the major functions: hiring, purchasing, workshop, transport, costing, administration/personnel, accounting. Within these functions the costs may be further collected into cost centres. For example, each item of equipment maintained by the workshop may be given a code number which may represent the cost centre of one or more similar items. The costs recorded for each cost function or centre may be subdivided into elements such as materials, labour and expenses:

1 Material cost: consumables and spares.
2 Labour cost: wages and salaries of the employees.
3 Expenses: depreciation of plant and equipment, repairs, administration, services provided, water and electricity.

The costs of materials, labour and expenses which can be clearly allocated to a cost centre are called *direct costs*, and usually vary with the volume of production or equipment usage. *Indirect costs* are those materials, labour and expenses which cannot be directly identified to the cost centre, but which provide some function or service, such as a computer or the rent of the firm's offices and works. Indirect costs are thus apportioned between the cost centres, and are usually referred to as overhead costs.

Indirect costs are mostly fixed costs such as staff salaries, rent and rates, insurances, office equipment, maintenance tools and machines, which remain constant irrespective of the volume of trading or work done. A direct or variable overhead is one which varies in cost with the volume of production, such as electricity.

Costing

While budgets are prepared from predetermined costs, because of short-term changes in business performance it is essential that the actual costs incurred be continuously monitored and compared with budgeted costs in order that changes may be implemented. The difference between the actual cost and the predetermined cost is called a variance. A costing system should be updated regularly on a weekly basis, and the variances calculated for each function, department or cost centre. The procedure may also include analysis of the variances incurred by the individual items of plant in the fleet.

A note of caution with regard to the budgets, particularly departmental budgets, is advised. It is useful to compare the actual result with the value for the same month or week of the previous year. Astute managers can be adept at 'hiding' behind a 'camouflaged' budget which may have fortuitously arisen through changed circumstances, for example, reduction in the assumed rate of inflation, manipulation of the figures, etc. Managers' performance should therefore be measured against both the budget and the previous year's results.

Control of the workshop budget

Example 14.1

The annual budget for a plant workshop is £100,000 (see Table 14.2). This figure is based upon the size of the fleet and the estimated hours that the fleet will be operated during the year.

The budgeted direct costs of the department consist of the purchase of consumable materials and spare parts and labour such as fitters, mechanics, etc., and these total £80,000. Budgeted indirect costs include staff salaries, rent, rates, insurances, general administration charges, depreciation of workshop equipment and power. These amount to £20,000 and are a fixed charge.

At the end of the year business activity had been lower than expected and the hours operated by the fleet were 10 per cent fewer than initial estimates. The actual direct costs of the workshop over this period were £70,000 and the actual cost of overheads was £21,000. The position is provided in Table 14.5.

Thus, although the level of activity anticipated at the beginning had not been realised, the works department had managed to maintain a favourable variance on direct costs. In this example only overheads produced an unfavourable variance and a more detailed analysis of costs should reveal the reasons, such as excess secretarial staff. However,

Table 14.5 Control of the workshop budget

	Planned budgeted costs for this date (£)	*Adjusted budgeted costs of work done (£)*	*Actual cost (£)*	*Variance (£)*
Direct costs	80,000	72,000	70,000	+2,000
Overheads	20,000	20,000	21,000	−1,000
Total	100,000	92,000	91,000	+1,000

because the volume of business had not reached the level anticipated, the *budgeted profit* for the enterprise also would not be fully recovered and the costs of the fixed overheads would have to be met from the reduced profits. For a plant hire business such consequences could be particularly severe, as much of the business costs are generated as fixed overheads. For a comprehensive review of performance, therefore, the sales variance should be included in the analysis.

Sales variance

Example 14.2

The budgeted hire revenue from equipment over the coming 12 months is £400,000. Profit, overheads and depreciation are set at 10 per cent, 20 per cent and 20 per cent respectively, of revenue. However, the actual revenue was only £350,000. The variances recorded for direct and indirect costs (overheads) were, respectively, + £3,000 and + £1,000. In addition, several items were written off prematurely, leading to a negative variance on equipment depreciation of £1,000.

Analysis of variance

Sales variance	=	−£25,000	(i.e. 50% of £400,000 − £350,000)
Direct costs variance	=	+£3,000	
Indirect costs variance	=	+£1,000	
Depreciation variance	=	−£1,000	
Actual shortfall on profits	=	£22,000	(i.e. £18,000 profit as compared to £40,000 budgeted profit).

It can be seen that the profit and overhead is under-recovered by £25,000 and only the collective economies made by the various departments reduced the magnitude of the shortfall to £22,000.

Other useful management information

Equipment utilisation reports

The sales variance may also be calculated for each group or item of equipment, to provide management with a regular update of the effects of changes in the levels of utilisation and hire rates. The utilisation report is produced monthly and is divided into two sections, representing equipment for hire and non-operated plant – that is, equipment hired without an operator. Each division is subdivided into individual cost centres or like machines – for example, scrapers, bulldozers, crawler cranes, etc. – and the utilisation and price variance *on sales* for each item in a cost centre is produced in the final report as shown below.

Analysis of variance

1 *Utilisation variance* is the financial effect of using the equipment either more or less than those hours budgeted. Thus:

  Utilisation variance = (actual hours × budgeted hire rate) − (budgeted hours × budgeted hire rate)

2 *Price variance* is the financial effect of charging more or less than the budgeted hire rate. Thus:

Price variance = (actual hours × actual hire rate) − (actual hours × budgeted hire rate)

The sum of the utilisation and price variances multiplied by the profit and overhead margin is the true sales variance on the item.

Example 14.3

During the month an item of equipment was hired out for 280 hours at a hire rate of £12 per hour. The budget anticipated only 200 hours of work at a hire rate of £13 per hour. Calculate the utilisation and price variances. The profit (and overhead) is set at 10 per cent of the hire rate.

1 Utilisation variance on sales

$= (280 \times 13) - (200 \times 13) = 80 \times 13 = +£1{,}040$

(favourable)

2 Price variance on sales

$= (280 \times 12) - (280 \times 13) = 280 \times -1 = -£280$

(unfavourable)

Sales variance = (actual revenue − budgeted revenue) × profit and overhead margin

$= (280 \times 12 - 200 \times 13) \times 0.1 = +£760.$

The effect of operating the equipment 80 hours more than planned increased the revenue by £1,040, but reduce to £760 because the hire rate was less favourable than budgeted. The variances should be cumulatively totalled for each month, to present a comprehensive record, which together with records collected of the hours operated for the machine, provides an indication of the competitiveness and excessive use or otherwise of the item.

These two variances signify to the management the popular and competitive equipment items to operate, and the information is therefore valuable in deciding upon purchases and disposals.

Equipment cost report

A positive utilisation variance would usually be associated with increased maintenance, affecting direct costs and possibly indirect costs also. Therefore, the budgetary and cost control system recommended for departments should be installed for individual equipment items, and actual costs and revenues should be recorded on a regular basis. In this way all the variances may be monitored and the consequences on profitability quickly recognised.

Example 14.4

During a six-month period £4,000 revenue was received for a piece of equipment. The budget anticipated £5,000. Direct costs and indirect costs (overheads) from operating the various departments of the enterprise were budgeted to the item at £1,500 and £2,000 respectively, with profit at 10 per cent of turnover, and £1,000 was provided for depreciation. Actual direct costs recorded during the period were £1,250 and the actual overhead incurred was £1,950. Calculate the variances and percentage return on sales.

Analysis of variance

Budgeted profit (10% of £5,000)	= £500
Sales variance (£1,000 × 70%)	= −£700
Direct cost variance	
$(£1{,}500 \times \frac{4{,}000}{5{,}000} - £1{,}250)$	= −£50
Overhead variance	= +£50
Depreciation variance	= £0
Total variance	= −£700
Actual profit on turnover	$= \frac{-200}{4{,}000} = -5.0\%$ (loss)

Note: Sales variance multiplier of 70 per cent is derived from profit, overhead and depreciation – that is, (£500 + £2,000 + £1,000)/£5,000.

A comprehensive record of costs, current written-down value, revenue and profitability for each item may be used in conjunction with the utilisation report, to purchase and dispose of items at prices which are economic for the company. For many firms this type of information is monitored on an asset register of the equipment holdings.

Marginal costing

Marginal costs are those costs arising directly from the production process, which for an equipment hire business would be largely those costs connected with maintenance and servicing of the equipment. They therefore vary directly with the hiring activity. Fixed costs, arising from the establishment charges, fluctuate very little with hiring levels. The purpose of the marginal costing method is to calculate the contribution made by each item of equipment for hire towards the fixed costs and profit of the business. For an example of marginal costing refer to Table 14.6.

This technique can be used to advantage during a short-term period when the market demand is low and hire rates need to be keen to attract custom, the contention being that any hire rate revenue which exceeds the marginal costs makes a contribution towards the fixed costs. However, such a pricing policy should be considered only during a short and difficult period, since the endeavour must be to realise the budgeted profit for each item over

Table 14.6 Marginal costing

		Plant item, £00s (weekly)				
		A	*B*	*C*	*D*	*E*
Hire revenue		10	9	5	15	10
Labour costs	Marginal costs	3	2	2	5	2
Material costs	Marginal costs	3	2	1	5	3
Expenses	Marginal costs	2	2	1	3	2
Contribution*		2	3	1	2	3
Contribution per £1 revenue		20p	33p	20p	13p	30p

Note
* Contribution = overheads + profit.

the 12-months periods. Thus, for plant item C, for example, a hire rate exceeding £400 per week will contribute to the fixed overhead, which may be a better alternative than leaving the machine idle. Conversely, the method gives a clear indication that the firm should be directing its sales effort on items B and E, as these machines can obtain favourable hire rates and give the best contribution towards fixed costs.

Chapter 15

Cash flow

Introduction

Bankruptcy or voluntary liquidation is caused not only by a lack of cash, but also by an inability to raise cash in the form of loans or credit to meet immediate commitments, because creditors, investors and possibly lenders of money – usually the banks – have lost confidence in the business and are not convinced that the company can continue to trade in a profitable and viable way. This loss of confidence is an important factor because it is the existence of such confidence that permits overdrafts to be obtained and normal trade credit to be received, and a loss in confidence would result in existing creditors pressing harder for payment. Trade credit is an important factor in determining most companies' short-term cash requirements and should trade credit be withheld, the short-term cash requirements increase significantly. The withholding of trade credit simply means that the suppliers to a company demand cash on delivery rather than invoicing, say, at the end of the month and requiring payment one month later.

Most of Chapter 6 was concerned with profitability measured in terms of return on capital, and much of Chapter 16, dealing with financial management, will be concerned with determining the company's profit and distributing that profit. Profit, seen simply as the difference between revenue and cost, is a common measure of a company's well-being, and the derived ratios of profit/turnover and profit/capital employed are useful indicators of the company's performance, but these are derived from measuring profit. Undoubtedly, a company with a good profit is likely also to have good profitability (i.e. profit as a proportion of capital employed) and, in turn, be in a good state to avoid liquidity problems. But the company's liquidity also needs to be monitored and managed. Although a company may be profitable, it may have liquidity difficulties. An unexpected demand for payment might not be possible to be met and may cause significant difficulties, if not bankruptcy.

The more detailed monitoring and managing of a company's cash flow can be seen as two related and integrated but different aspects. One is the cash required for normal trading operations and the other is the cash required for acquisitions, less disposals.

The cash required for normal trading operations is controlled by sales fluctuations, trade credit (creditors, less debtors), stocks, work in progress and perhaps value added tax (VAT). In equipment hire, sales fluctuations are manifest by the utilisation of equipment. Equipment is idle in a sales slump, but all equipment is highly utilised and more new equipment is required in a sales boom. Stocks in equipment hire are less important than in manufacturing and normally only represent spares, repair materials, fuel and oil. Also, equipment hire companies differ from manufacturing companies in that there is no manufacturing process absorbing manpower and materials and locking up cash. The nearest equivalent to 'work in progress' in equipment hire is equipment on hire for which the invoices have not been issued, or perhaps equipment under repair.

The other aspect is the cash required for the provision of the company's capital assets. This is particularly important to equipment hire companies, since as much as 50 per cent of sales turnover may be used in meeting the costs of asset ownership. The variables that control this are the purchase and disposal of capital assets, the method of acquisition which, in turn, controls the methods of payment (e.g. purchase, hire purchase or lease) and the company's ability to utilise capital allowances. In addition to these factors, the company's cash flow is also affected by interest and other bank charges, corporation tax and dividends. Corporation tax is important to the cash flow, as at the time of writing, capital allowances of 25 per cent of the purchase price written down are allowed in acquiring an asset, such as an item of construction equipment. Therefore, the disposal and acquisition of equipment is significant when determining the corporation tax due and, in turn, the acquisition of equipment and payment of corporation tax are significant to the company's cash requirements. Dividends are also significant to the company's cash position, but within the control of the company's directors.

The following description of cash flow problems reflects the two main aspects of cash flow: (1) the cash flow resulting from the normal month-to-month trading operations and (2) the cash flows resulting from the acquisition and disposal of capital assets. These will be considered separately.

Cash flows from normal trading operations

Trade credit, sales fluctuations and stocks

The factors that affect the short-term cash requirements in equipment hire are trade credit, sales fluctuations and stocks. The effects of these are best illustrated by considering an example which shows the effects of these factors one at a time. In manufacturing, the additional factor of 'work in progress' is important, but as there is no process in equipment hire that can be classed as work in progress, it is not significant and is not included in this example.

The monthly net cash flow stated in Example 15.1 is the difference between the cash outgoings and incomings in that month. The monthly contribution is the revenue that will be derived from the sales (hired-out equipment) during the month, less the direct costs in supporting the sales for labour, spare parts and consumables. The word 'contribution' is used because this money will be used to meet overhead costs and ownership costs and therefore cannot yet be described as 'profit'. All transactions in this example are exclusive of value added tax, which is referred to separately.

Example 15.1

An equipment hire company initially holds five items of equipment which are hired out at the rate of £1,000 per month. Each month's maintenance on each item of equipment uses £200 in spare parts and £100 in consumables such as fuel, oil and grease. The fuel costs for running the equipment are the responsibility of the hirer. The workshop providing the maintenance support has sufficient labour to maintain five items of equipment in working order each month. The labour cost of the workshop is £750 per month. If all the five items of equipment were on hire and all the transactions were in cash, the monthly cash flow would be as follows:

Month 1 (with all transactions in cash)

	Cash out (£)	*Cash in (£)*
Sales (5 items of equipment on hire)		5,000
Workshop labour	750	
Purchases		
Spares for 5 items of equipment for 1 month	1,000	
Consumables for 5 items of equipment for 1 month	500	
Totals	2,250	5,000
Net cash flow for month 1	+£2,750	
Contribution to company for month 1	+£2,750	

The contribution is defined as the revenue derived from the hire sales, less the direct costs incurred in supporting these sales, such as labour, spares and consumables. The company profit will be the contribution, less the ownership and overhead costs. If the hirers were given trade credit of one month to pay, then the cash flow for month one would be as follows.

Month 1 (with trade credit for hirers)

	Cash out (£)	*Cash in (£)*
Sales (5 items of equipment on hire)		nil
Workshop labour	750	
Purchases		
Spares for 5 items of equipment for 1 month	1,000	
Consumables for 5 items of equipment for 1 month	500	
Totals	2,250	nil
Net cash flow for month 1	−£2,250	
Contribution to company for month 1	+£2,750	

The company has the same contribution but the effect of trade credit to the hirers produces a net cash flow of −£2,250. If the suppliers of spares and consumables also offered one month trade credit but the labour was paid weekly, the cash flows would be as follows.

Month 1 (with trade credit for hirers and from suppliers)

	Cash out (£)	*Cash in (£)*
Sales (5 items of equipment on hire)		nil
Workshop labour	750	
Purchases		
Spares for 5 items of equipment for 1 month	nil	
Consumables for 5 items of equipment for 1 month	nil	
Totals	750	nil
Net cash flow for month 1	−£750	
Contribution to company for month 1	+£2,750	

Thus, although the contribution generated in this month was again £2,750, the cash resources required were reduced to −£750 by the availability of trade credit from suppliers. Assuming that this is the normal trading experience, the cash flows in the following month, month 2, will be different from those in month 1, even though the same level of sales (i.e. hired-out equipment) is achieved. This is because the cash flows from month 1 sales and purchases will be present.

Month 2 (with trade credit for hirers and from suppliers)

	Cash out (£)	*Cash in (£)*
Sales (5 items of equipment on hire)		nil
Workshop labour	750	
Purchases		
Spares for 5 items of equipment for 1 month	nil	
Consumables for 5 items of equipment for 1 month	nil	
Revenue from previous month's sales		5,000
Payments for previous month's purchases		
Spares	1,000	
Consumables	500	
Totals	2,250	5,000
Net cash flow for month 2	+£2,750	
Contribution to company for month 2	+£2,750	

In month 2 the cash flow is a net inflow of +£2,750, arising from the revenue from month 1 sales. The contribution earned in month 2 was again £2,750 and the cash flows and contributions for the two months are:

	Cash flow (£)	*Contribution (£)*
Month 1	−750	+2,750
Month 2	+2,750	+2,750

Provided that the five items of equipment are hired out and maintained each month, the cash flows and contributions will continue as in month 2. If in months 3 and 4 only two items of equipment are hired out, the pattern is disturbed, as follows.

Month 3 (with trade credit for hirers and from suppliers)

	Cash out (£)	*Cash in (£)*
Sales (2 items of equipment on hire)		nil
Workshop labour	750	
Purchases		
Spares for 2 items of equipment for 1 month	nil	
Consumables for 2 items of equipment for 1 month	nil	
Revenue from previous month's sales		5,000
Payments for previous month's purchases		
Spares	1,000	
Consumables	500	
Totals	2,250	5,000
Net cash flow for month 3	+£2,750	
Contribution to company for month 3	+£650	
Number of idle items of equipment	3	

(The contribution is calculated as the revenue from two items of equipment of £2,000, less the workshop labour of £750, and the spares for two items of equipment of £400 and consumables for two items of equipment of £200.)

Although the sales in month 3 were reduced, the effect was not shown immediately on the cash flow because of trade credit. The reduction in contribution generated in the month is aggravated by the fact that the labour costs could not be reduced in a similar way, as the purchases were reduced to levels compatible with the reduced sales. A serious problem of reduced sales not shown here but which will be dealt with later is that the ownership costs of the idle equipment still have to be met from the reduced contribution. The effects of reduced sales revenue come through in month 4.

Month 4 (with trade credit for hirers and from suppliers)

	Cash out (£)	*Cash in (£)*
Sales (2 items of equipment on hire)		nil
Workshop labour	750	
Purchases		
Spares for 2 items of equipment for 1 month	nil	
Consumables for 2 items of equipment for 1 month	nil	
Revenue from previous month's sales		2,000
Payments for previous month's purchases		
Spares	400	
Consumables	200	
Totals	1,350	2,000
Net cash flow for month 4	+£650	
Contribution to company for month 4	+£650	
Number of idle items of equipment	3	

Thus, the smaller sales revenue from month 3 has reduced the net cash inflow from £2,750 to £650.

If in month 5 the sales improve to four items of equipment on hire, the improved cash flows again take a further month to come through, as months 5 and 6 illustrate.

Month 5 (with trade credit for hirers and from suppliers)

	Cash out (£)	*Cash in (£)*
Sales (4 items of equipment on hire)		nil
Workshop labour	750	
Purchases:		
Spares for 4 items of equipment for 1 month	nil	
Consumables for 4 items of equipment for 1 month	nil	
Revenue from previous month's sales		2,000
Payments for previous month's purchases		
Spares	400	
Consumables	200	
Totals	1,350	2,000
Net cash flow for month 5	+£650	
Contribution to company for month 5	+£2,050	
(Contribution = sales − workshop labour − spare − consumables)		
Number of idle items of equipment	1	

Month 6 (with trade credit for hirers and from suppliers)

	Cash out (£)	*Cash in (£)*
Sales (4 items of equipment on hire)		nil
Workshop labour	750	
Purchases		
Spares for 4 items of equipment for 1 month	nil	
Consumables for 4 items of equipment for 1 month	nil	
Revenue from previous month's sales		4,000
Payments for previous month's purchases:		
Spares	800	
Consumables	400	
Totals	1,950	4,000
Net cash flow for month 6	+£2,050	
Contribution to company for month 6	+£2,050	
(Contribution = sales − workshop labour − spare − consumables)		
Number of idle items of equipment	1	

The cash flow in month 6 reflects the increase in sales experienced in month 5.
The cash flow and contribution for the first six months of this venture are therefore:

Month	*1*	*2*	*3*	*4*	*5*	*6*
Net cash flows	−£750	+£2,750	+£2,750	+£650	+£650	+£2,050
Contribution	+£2,750	+£2,750	+£650	+£650	+£2,050	+£2,050

It is clear from this table that the contribution more closely reflects the fluctuating sales, whereas the effects of trade credit produce different cash flows.

So far the company has been purchasing just enough spares and consumables for each month's operations, but if it wished to build up stocks of spares and consumables, then the cash flows would be reduced in order to fund the stocks. Months 7 and 8 are examples of building up stocks to meet a future sales boom.

Month 7 (with trade credit for hirers and from suppliers)

	Cash out (£)	*Cash in (£)*
Sales (4 items of equipment on hire)		nil
Workshop labour	750	
Purchases		
Spares for 8 items of equipment for 1 month	nil	
Consumables for 8 items of equipment for 1 month	nil	
Revenue from previous month's sales		4,000
Payments for previous month's purchases		
Spares	800	
Consumables	400	
Totals	1,950	4,000
Net cash flow for month 7	+£2,050	
Contribution to company for month 7	+£2,050	
Number of idle items of equipment	1	
Stocks		
Spares for 4 items of equipment for 1 month	£800	
Consumables for 4 items of equipment for 1 month	£400	

In month 7 spares and consumables for eight items of equipment were purchased, although only four were required to support the sales in the month. The four purchased in excess are held in stocks, as shown. Contribution is calculated on the revenue and cost of the sales.

Month 8 (with trade credit for hirers and from suppliers)

	Cash out (£)	*Cash in (£)*
Sales (4 items of equipment on hire)		nil
Workshop labour	750	
Purchases		
Spares for 8 items of equipment for 1 month	nil	
Consumables for 8 items of equipment for 1 month	nil	
Revenue from previous month's sales		4,000
Payments for previous month's purchases		
Spares	1,600	
Consumables	800	
Totals	3,150	4,000
Net cash flow for month 8	+£850	
Contribution to company for month 8	+£2,050	
Number of idle items of equipment	1	
Stocks:		
Spares for 8 items of equipment for 1 month	£1,600	
Consumables for 8 items of equipment for 1 month	£800	

Thus, the build-up of stocks has reduced the cash flow to +£850.

To meet the expected sales boom, the company increases the total holding to eight items of equipment. In order to service this, additional labour is required, which brings the workshop labour cost to £1,800.

Month 9 (with trade credit for hirers and from suppliers)

	Cash out (£)	*Cash in (£)*
Sales (8 items of equipment on hire)		nil
Workshop labour	1,800	
Purchases		
Spares for 8 items of equipment for 1 month	nil	
Consumables for 8 items of equipment for 1 month	nil	
Revenue from previous month's sales		4,000
Payments for previous month's purchases		
Spares	1,600	
Consumables	800	
Totals	4,200	4,000
Net cash flow for month 9	−£200	
Contribution to company for month 9	+£3,800	
(Contribution = sales − workshop labour − spares − consumables)		
Number of idle items of equipment	nil	
Stocks		
Spares for 8 items of equipment for 1 month	£1,600	
Consumables for 8 items of equipment for 1 month	£800	

The increase in sales has led to a negative cash flow, because the sales have led to increased trade credit which will not show as cash inflows until the next month and the increased workshop costs had to be met immediately. Also, the level of purchases of spares and consumables was high in month 8 and cash must be found to fund the stocks. Cash flow for month 10 will improve as the increased revenues come through. Such rapid expansion could, if continued, lead to overtrading in which the company's cash resources would not be sufficient to support the expansion.

Month 10 (with trade credit for hirers and from suppliers)

	Cash out (£)	*Cash in (£)*
Sales (8 items of equipment on hire)		nil
Workshop labour	1,800	
Purchases		
Spares for 8 items of equipment for 1 month	nil	
Consumables for 8 items of equipment for 1 month	nil	
Revenue from previous month's sales		8,000
Payments for previous month's purchases		
Spares	1,600	
Consumables	800	
Totals	4,200	8,000
Net cash flow for month 10	+£3,800	
Contribution to company for month 10	+£3,800	
Number of idle items of equipment	nil	
Stocks		
Spares for 8 items of equipment for 1 month	£1,600	
Consumables for 8 items of equipment for 1 month	£800	

The cash flow in month 10 now reflects the higher revenue of the sales in month 9. If in month 11 some of the stock held is used to maintain equipment rather than buying new spares and consumables, the cash flow would further improve in month 12.

Month 11 (with trade credit for hirers and from suppliers)

	Cash out (£)	*Cash in (£)*
Sales (8 items of equipment on hire)		nil
Workshop labour	1,800	
Purchases		
Spares for 4 items of equipment for 1 month	nil	
Consumables for 4 items of equipment for 1 month	nil	
Revenue from previous month's sales		8,000
Payments for previous month's purchases		
Spares	1,600	
Consumables	800	
Totals	4,200	8,000
Net cash flow for month 11	+£3,800	
Contribution to company for month 11	+£3,800	
Number of idle items of equipment	nil	
Stocks		
Spares for 4 items of equipment for 1 month	£800	
Consumables for 4 items of equipment for 1 month	£400	

Although eight items of equipment were on hire, spares and consumables were only purchased for four. The remaining required spares and consumables were taken from stock and the value of the stock was reduced.

Month 12 (with trade credit for hirers and from suppliers)

	Cash out (£)	Cash in (£)
Sales (4 items of equipment on hire)		nil
Workshop labour	1,800	
Purchases		
Spares for 4 items of equipment for 1 month	nil	
Consumables for 4 items of equipment for 1 month	nil	
Revenue from previous month's sales		8,000
Payments for previous month's purchases		
Spares	800	
Consumables	400	
Totals	3,000	8,000
Net cash flow for month 12	+£5,000	
Contribution to company for month 12	+£1,000	
Number of idle items of equipment	4	
Stocks		
Spares for 4 items of equipment for 1 month	£800	
Consumables for 4 items of equipment for 1 month	£400	

The cash flow increases to £5,000 in month 12 when the revenues are £8,000 due to the high sales in month 11 and payments for purchases are only £1,200 because half the spares and consumables used in month 11 were drawn from stock rather than purchased. This is known as de-stocking. The cash flows and contributions are summarised in Table 15.1.
Summary of the difference between cash and contribution in Table 15.1:

- The cumulative contribution for the 12 months is £27,400.
- The cumulative cash flow for the 12 months is £23,400.
- The difference between these is due to the trade credit and stocks.

Trade credit	
Creditors	
Outstanding credit received from suppliers	
Spares	£800
Consumables	£400
	£1,200
Debtors	
Outstanding credit given to hirers	
4 items of equipment on hire	£4,000
Difference between debtors and creditors	£2,800
Stocks	
Spares	£800
Consumables	£400
Total	£1,200

Table 15.1 Summary of cash flows, contributions and stocks for 12 months

Month	*1*	*2*	*3*	*4*	*5*	*6*	*7*	*8*	*9*	*10*	*11*	*12*
Contribution (£)	2,750	2,750	650	650	2,050	2,050	2,050	2,050	3,800	3,800	3,800	1,000
Net cash flow (£)	−750	2,750	2,750	650	650	2,050	2,050	850	−200	3,800	3,800	5,000
Stocks (£)	—	—	—	—	—	—	1,200	2,400	2,400	2,400	1,200	1,200
Reasons for cash flow fluctuations	Trade credit given to hirers and received from suppliers			Falling sales in month 3		Increased sales in month 5		Increase in stocks in month 7	Increase in sales leading to increasing trade credit and increase in workshop labour costs to support new sales and stocking in month 8		De-stocking in month 11 while sustaining high sales levels in month 11	

Thus, the summary of trade credit and stocks is

Trade credit	£2,800
Stocks	£1,200
Total	£4,000

Of the £27,400 of contribution earned, the company has: £23,400 in cash (less overheads and ownership costs, etc.); £2,800 in debtors less creditors; and £1,200 in stocks.

The cash flows calculated in Table 15.1 were calculated on the basis of trading operations and require further adjustment before they represent the company cash flow. These other adjustments are head office or overhead expenses, the acquiring and disposal of capital assets, value added tax, corporation tax and dividends.

Overheads

Adjusting the cash flows and profits for overheads requires a projection of the head office overheads. Some of these expenses, such as salaries, will be monthly; others, such as telephone, electricity and rentals, may be monthly or quarterly; while business rates may be half-yearly. Table 15.2 shows the original contribution and cash flows from Table 15.1 with the overhead adjustment included.

The overhead projections in Table 15.2 are as shown. These have been stated in the month in which they occur, and the same figure in any month has been deducted from the contribution as well as the cash flow. The difference between the cumulative contribution of £27,400 and the cumulative cash of £23,400 is due to stocks and the difference between debtors and creditors, as previously explained. The effect of overheads is simply to reduce the contribution and cash available.

Table 15.2 Summary of profit and cash flows for 12 months with overheads included

Month	*Contribution before overheads deducted (Table 15.1) (£)*	*Cash flow before overheads deducted (Table 15.1) (£)*	*Overheads (£)*	*Contribution less overheads (£)*	*Cash flow less overheads (£)*
1	2,750	−750	150	2,600	−900
2	2,750	2,750	10	2,740	2,740
3	650	2,750	40	610	2,710
4	650	650	10	640	640
5	2,050	650	10	2,040	640
6	2,050	2,050	190	1,860	1,860
7	2,050	2,050	10	2,040	2,040
8	2,050	850	10	2,040	840
9	3,800	−200	40	3,760	−240
10	3,800	3,800	10	3,790	3,790
11	3,800	3,800	10	3,790	3,790
12	1,000	5,000	40	960	4,960
Cumulative values	27,400	23,400	530	26,870	22,870

Value added tax

A company collects value added tax (VAT) on receipt of payment from customers. The company also pays VAT on the payment of suppliers for goods received. The difference is calculated and either paid to or received from the Customs and Excise office quarterly. Depending on the difference in value between the amount of goods bought and sold in a quarter, the VAT can result in an outflow or an inflow. If the VAT is owed by the company, the collection of VAT has been acting as a source of short-term funds. If a VAT refund is due to the company, then the company has been funding the difference, and the effect of VAT on the company's cash flow should be calculated.

Cash flow forecasts for trading operations in equipment hire

Trade credit

Equipment hire companies give and receive credit. The credit given to customers is typically one month, so that any equipment hired this month will have an invoice issued at the time of supply and payment will be expected in, say, 30 days. In preparing a cash flow forecast from the sales forecast, the cash in, or revenue, can be derived by inserting the appropriate time shifts between the month in which the sales occur and the month in which the cash is received. In deriving revenue, it is usual to assume that not all the customers will pay on the due date and that some will default. An analysis of the timing of receipts from customers will indicate which assumed time shifts are appropriate. Representative figures would show that 70 per cent of customers pay within one month, 25 per cent within two months and 5 per cent within three months. It may be that payment is never received on some invoices, and this should be taken into account in the cash forecast if this proportion is significant. Thus, a sales forecast and derived cash in or revenue would be as illustrated in Table 15.3.

Similarly, credit would be received from suppliers of spare parts and consumables, and from the forecast of required spare parts, repair materials and consumables the appropriate time shifts would be used to determine the cash out. Labour and operatives would be paid weekly and the staff monthly.

Stocks and work in progress

As an equipment hire operation is not involved in manufacture, there is no substantial purchase of raw materials leading to stocks, nor is there any substantial work in progress. Stocks are mainly confined to spare parts and repair materials.

Table 15.3 Sales forecast and derived cash revenue

Month	*1*	*2*	*3*	*4*	*5*	*6*	*7*
Sales forecast (£)	10,000	15,000	18,000	12,000	etc.	etc.	etc.
1 month delay (70%) (£)		7,000	10,500	12,600	8,400	etc.	etc.
2 month delay (25%) (£)			2,500	3,750	4,500	3,000	etc.
3 month delay (5%) (£)				500	750	900	600
Cash in (£)		7,000	13,000	16,850	13,650	3,900	600

Sales fluctuations and overheads

As with any type of business, the equipment hire industry is subject to sales fluctuations. These, like the earlier examples, lead to increases in trade credit and can lead to overtrading in a sustained expansion. Stocking and destocking of spare parts and consumables takes place within the cycle of sales fluctuations.

The bouyancy of the equipment hire industry is correlated with the output of the whole of the industry. Major changes in the level of output can lead to problems of overtrading on the rising markets, and later when lower levels of industrial activity occur, companies are left with disproportionate overheads. The organisation is faced with cutting overheads as the market shrinks and the company's turnover declines, but the equipment hire company has another, more difficult, problem. This problem is that the equipment hire company has capital assets in the form of off-highway plant and equipment which, if they become

Table 15.4 A tabular form for the construction of a cash flow forecast for trading operations

Month	*1*	*2*	*3*	*4*	*5*
Sales forecast					
1 month delay (70%)					
2 month delay (25%)					
3 month delay (5%)					
Revenue or cash in					
Workshop					
Goods purchased					
cash out					
Labour payroll					
cash out					
Staff payroll					
cash out					
Overheads (rent, rates, phone, etc.)					
cash out					
Total cash out for workshop					
Transport department					
Goods purchased					
cash out					
Labour payroll					
cash out					
Staff payroll					
cash out					
Overheads					
cash out					
Total cash out for transport department					
A cash-out flow for each section or main head of account in budget would be prepared					
Total cash out from all sections					
Total cash in from sales					
Net cash flow					
Cumulative cash flow					

under-utilised or hired out at an uneconomic rate, or both, quickly cease to give a return on capital. In such a situation companies would try to dispose of their equipment, but in a market slump this may not be possible. This is the basic risk in all equipment hire operations.

Forecasts

A short-term cash flow forecast is derived from the sales forecast and the aggregate cash requirements of all the various heads of account, divided into purchasing supplies, goods, labour, staff and overheads. The heads of account are the same as those used in the budget – namely workshop, transport and the various service departments of purchasing, administration, costing, hiring and finance. A tabular flow forecast may be constructed as illustrated in Table 15.4.

Cash flows for the purchasing and acquisition of capital assets

One feature of equipment hire companies that distinguishes them from general contractors is the high proportion of funds locked up in their capital assets. As much as 40–50 per cent of the sales revenue may be dissipated on the ownership costs of the company's assets. The cash flow for the normal month to month trading as in Tables 15.2 and 15.4 should be combined with the company's capital expenditure budget, which includes disposals as well as purchases, to produce the cash flow statement.

As the capital expenditure budget is likely to produce the larger cash flows and will dominate the company cash flow, it is not unusual for the capital expenditure budget to be prepared first.

The method of acquisition determines the cash flows for individual items and Chapter 6 reviewed the various methods of acquisition: outright purchase, credit purchase, hire purchase or lease. All have different cash flow implications, and the cash flows of these various methods of acquisition can be summarised as follows:

Method of acquisition	*Cash flow*
Outright purchase	Single large down payment
Credit purchase	Deposit plus regular payments
Hire purchase	Deposit plus regular hire charges
Lease	No deposit but regular lease payments

Thus, leasing is the least difficult to provide for, because it should be paid from revenue. Hire purchase or some credit arrangement is the next easiest, since a down payment is required and the remaining charges can be met from revenue. However, the down payment may be borrowed and the loan repayments met from revenue. Similarly, outright purchase requires a large single payment of cash, but all of this or part of it can be borrowed and payments can be met from revenue.

Providing a cash flow statement that includes the effects of acquisition and purchasing requires a prediction of the proceeds received from disposals together with a forecast of the expenditure on acquisition divided into down payments, hire purchase or loan repayments, interest charges and lease payments. The following example has been chosen so that it may be included with the cash flow calculated in Table 15.2.

Example 15.2

Of the five items of equipment initially acquired by the company, three were leased and two were bought on hire purchase. The outright purchase price of the plant items is £5,000. The lease payments are £270 each per month for 24 months. The HP deposit on one plant item is £1,000 and the monthly payments are £385 for 12 months. In month 8, when the plant holding was increased to eight units, this was done by selling one of the plant items bought by hire purchase for £4,000, paying the HP company £1,320, representing the outstanding HP payments, less the interest that would be charged for the remaining period of four months. The fleet was increased by a further four leased items at a cost of £270 per month each. The cash flows for these acquisitions and disposals are shown in Table 15.5. Cash flow for acquisitions and disposal is now added to the cash flows for normal trading from Table 15.2 and is shown in Table 15.6.

Value added tax has previously been described. Interest charges will be due if the negative cash flow is funded from loans or an overdraft, and corporation tax will be due after all costs and interest have been deducted from sales revenue. The time lags associated with corporation tax will vary from about nine months to perhaps 24 months, and the payments due would be for the previous trading years. Dividends would be paid out on the decision of the board of directors and are determined by the company's cash position, among other factors.

To complete the equipment hire company's cash flow forecast the analysis that was suggested in Table 15.4 needs to be extended as in Table 15.7.

Aspects that perhaps need further explanation are the recovery of the capital monies invested in equipment through hire charges and the implications of corporation tax.

Sales revenue, depreciation and corporation tax

In a hire company the sales revenue will be exclusively or predominantly made up of hire charges for hired equipment. These hire charges will have been set at the current going market rate but will have been chosen not only for market reasons, but also for economic ones. The 'economic' hire rate will include four elements: direct costs, indirect costs or overheads, ownership costs and profit. Chapter 7 dealt with such hire rate calculations and described the various methods of including the ownership costs. All of these are based on a concept of depreciation. This allowance for depreciation added into the hire rate is intended to ensure that the hire charges are large enough to recover the invested capital and an appropriate return on that invested capital. Thus, the monthly sales revenue includes this depreciation allowance. In the cash flow calculations undertaken in this chapter no attempt has been made to link the depreciation allowance included in the hire charges with the monies to be paid in ownership costs such as HP, loan repayments or lease payments. Although in calculating hire rates

$$\text{sales revenue} = \text{direct costs} + \text{indirect costs} + \text{depreciation} + \text{profit}$$

it is not adequate to say

$$\text{profit} = \text{sales revenue} - \text{direct costs} - \text{indirect costs} - \text{depreciation}$$

and that the monies available to meet ownership costs are depreciation plus retained profits.

Table 15.5 Cash flow for acquisition and disposal of equipment

Month	*1*	*2*	*3*	*4*	*5*	*6*	*7*	*8*	*9*	*10*	*11*	*12*
Equipment acquisition												
5 items, value (£)	25,000											
3 items by lease (£)	15,000											
2 items by HP (£)	10,000											
Lease payments (£)	810	810	810	810	810	810	810	810	810	810	810	810
HP deposit (£)	2,000											
HP payments (£)	770	770	770	770	770	770	770	770	385	385	385	385
Equipment disposal (£)								4,000				
HP settlement (£)								1,320				
Equipment acquisition												
4 items, value (£)								20,000				
4 items by lease (£)								20,000				
Lease payments (£)								1,080	1,080	1,080	1,080	1,080
Total cash out (£)	3,580	1,580	1,580	1,580	1,580	1,580	1,580	3,980	2,275	2,275	2,275	2,275
Total cash in (£)								4,000				
Net cash flow (£)	−3,580	−1,580	−1,580	−1,580	−1,580	−1,580	−1,580	+20	−2,275	−2,275	−2,275	−2,275

Table 15.6 Cash flow for normal trading from Table 15.2, plus cash flows associated with plant acquisition and disposal from Table 15.5

Month	*Net cash flow from Table 15.2 (normal trading) (£)*	*Net cash flow from Table 15.5 (acquisition and disposal) (£)*	*Sum or total net cash flow (£)*	*Cumulative cash flow (£)*	*Contribution, less overheads from Table 15.2, less net cash flow from acquisition and disposal (£)*
1	−900	−3,580	−4,480	−4,480	−980
2	2,740	−1,580	1,160	−3,320	1,160
3	2,710	−1,580	1,130	−2,190	−970
4	640	−1,580	−940	−3,130	−940
5	640	−1,580	−940	−4,070	460
6	1,860	−1,580	280	−3,790	280
7	2,040	−1,580	460	−3,330	460
8	840	+20	860	−2,470	2,060
9	−240	−2,275	−2,515	−4,985	1,485
10	3,790	−2,275	1,515	−3,470	1,515
11	3,790	−2,275	1,515	−1,955	1,515
12	4,960	−2,275	2,685	730	−1,315

Note
Adjustments to this cash flow that would render it a true company cash flow would be for VAT, interest, corporation tax and dividends.

Table 15.7 Headings for the cash flow forecast for acquisitions and disposals

Item	*Month*					
	1	*2*	*3*	*4*	*5*	*etc.*
Plant disposals						
Outright purchases						
Down payments						
Hire purchase payments						
Loan repayment						
Lease repayment						
Interest						
Value added tax						
Corporation tax						
Dividends						

Profit is not calculated in this way for tax purposes, as capital allowances are used to offset the cost of acquiring capital assets (see Chapter 16). Furthermore, the partitioning of the sales revenue in this way unrealistically suggests that the part of the revenue described as 'depreciation' is unnecessarily limited to be used in meeting ownership costs.

The reason that the sales revenue was left unpartitioned is that the funds can be deployed in any way to suit the company's operations. If the need is to meet the funding of debtors due to increasing sales or stocks or to buy a new item of equipment the surplus of revenue over cost can be deployed in the most advantageous way to suit the company. The company's need for cash is the criterion which dictates the use to which that cash available is put. The original method of calculating the hire charge which made allowances for depreciation was only a method of arriving at a realistic hire charge and has no influence on how the income is used. Thus, the cash flow analysis recommended in this chapter does not partition the revenue under the heading used to calculate the original hire charges.

The implications of corporation tax also need to be understood in the context of the depreciation element of sales revenue. If a company buys a capital asset for £1,000, UK tax legislation allows this to be offset against tax at the rate of 25 per cent written down and in that year tax is charged on the remaining profits. In the first year this amounts to £250, to be set against tax. In the subsequent year 25 per cent of the remaining capital sum of £750 (i.e. £187.50) is set against tax and so on. The company's internal depreciation included in the hire charge is unlikely to recover the capital in exactly the same way or at the same rate. It may be that the company recovers the capital over three years including one-third of the capital in each year. Thus, in the year of purchase, if the company's profits were £1,200 (i.e. sales revenue less costs), the taxable profit would be £950, less interest charges, which is £1,200, less the capital allowances. If, in the next year, the company's profits (sales revenue, less costs) were £1,300, the taxable profit would be £1,112.50, less interest charges. As explained in Chapter 6, the method of acquisition influences the capital allowances. The internal depreciation included by the company in their hire charges has no bearing on the calculation of taxable profit.

The corporation tax implications from the three methods of acquisition – outright purchase, hire purchase and leasing – can be summarised as follows:

Method of acquisition	*Implications for corporation tax*
Outright purchase	A capital allowance of 25 per cent written down of the purchase price can be set against profits.
Hire purchase	A capital allowance of 25 per cent written down of the purchase price can be set against profits. The interest element of the HP charge is deductable from revenue before tax.
Leasing	No capital allowance is available but lease payments are deductable from revenue before tax.
Interest charges	Interest charges on loans and overdrafts are deductable from revenue before tax.

Thus, the method of acquisition which may be chosen for cash flow reasons has a bearing on the corporation tax due, which, in turn, affects the company's cash flow. The benefit of capital allowances can be derived only if there are adequate profits against which the allowances can be set.

Cash flow management

This chapter has dealt with explanations of how and why cash flows vary and the need to forecast cash requirements. There has been an accompanying implication that the cash flow can be managed, which is so within limits. The variables of sales fluctuations, the amount of credit given, level of stocking, when to dispose of capital assets, when to acquire capital assets and by what method they should be acquired are all subject to some managerial control. However, there are limits.

Sales fluctuations

It is theoretically easy in an expanding market to control sales growth and, hence, the increase in debtors. However, increasing sales may be difficult to resist but are likely to be restricted if this means acquiring more capital assets. It is more difficult to improve sales in a declining market: cutting hire charges may be an option but this could lead to unprofitable trading. Thus, if a major slump in the construction industry occurs, some equipment hire companies will inevitably cease to trade. A slump in the construction industry is out of the control of equipment hire companies.

Trade credit

The amount of trade credit is set by the general trading conditions, and if a company sets shorter credit periods than competitors, the company may lose customers. Thus, trade credit is not wholly within the company's control. However, vigorous credit control can ensure that invoices are not allowed to remain unpaid for long periods beyond the normal credit given. A credit control system is a very important feature of a company's cash flow management. The company's credit controllers have some control, in that credit can always be refused if the customer is deemed unworthy of the risk.

With respect to the credit received from suppliers, it is unlikely that a company could extend the normal credit arrangements without losing discount and jeopardising confidence in the company. Confidence is important in obtaining credit, overdrafts and loans.

Stocking

The level of stocks of spare parts, repair materials and consumables is a matter for equipment companies to decide. The equation that is being balanced is the cost of holding such stock against the risk of equipment remaining idle while spare parts are sought.

Disposal of capital assets

The timing of disposals is within the control of company managers, but the capital raised by such disposals is controlled more by general market conditions than by the company manager's needs.

Acquisition of capital assets

The timing and the method of acquisition are both within the control of company managers, subject to delivery delays, etc.

Risk

Another need for cash that has not been considered so far is access to cash to cover risk. It is possible to plan for routine maintenance and, drawing on experience, it is possible to plan for repairs within reason, but it is not difficult to imagine situations where expensive repairs are required unexpectedly. The option of delaying repairs until cash becomes available is usually an undesirable situation, as this leaves the equipment item idle and ownership costs accrue whether the item is idle or not. Consequently, unexpected and expensive repairs often need to be undertaken immediately, and provision for such situations must be made. This implies the ready availability of cash for equipment not covered by an engineering insurance policy, and the ready availability of cash could suggest that cash may be idle and not working unless invested in short-term investments.

Cash, deficits and surplus

The cash flow illustrated in Table 15.6 showed that the monthly net cash flow in that example varied from −£4,480 to +£2,685. If this example were scaled up to a more realistic size, it would show that the company cash flow can swing from considerable deficits to substantial surplus. The company needs a source of funds to cover the deficits and a reasonable use for the surplus that will earn returns without leaving the cash idle. The traditional source of funds in excess of the company's own cash is overdrafts. The use of surplus cash includes short-term investments that can be quickly realised to meet cash requirements. In times of high inflation and high interest rates the contribution of such investments can be a source of considerable income to companies.

Chapter 16

Financial management

Introduction

An equipment hire or rental concern, like any other, usually has limited liability status, with either private or publicly quoted shares (PLC), depending upon its ability to secure a quotation on the Stock Exchange. A limited company, irrespective of its size, must file its annual financial accounts with the Registrar of Companies. To ensure that these give **a true and fair** view of the company's trading position, they are subjected to an annual audit by an independent firm of approved professional accountants. If the enterprise, on the other hand, is a subsidiary or department of a parent company, its trading position need only be included generally in the controlling company's consolidated accounts.

Beyond these legal requirements the financial accounts provide a basis for measuring the profit made by the company and its overall financial performance during the year past. In addition, the accounts give information on the investment policy, borrowing and other details of interest to shareholders (the owners), other investors and creditors. In the annual report, the main items presented are the Profit and Loss Account, the Balance Sheet and notes to the accounts.

The profit and loss account

The Profit and Loss Account is a statement of the company's total profit or loss resulting from trading during the year. Its main features are the revenues generated from sales together with the costs incurred in producing the sales. The difference between the two values represents the profit or loss. After deduction of interest charges on borrowed capital and corporate tax, the resulting surplus is used to provide a dividend to shareholders and/or for reinvestment in the company as retained earnings.

Example 16.1: Profit and loss account for the current year (including capital allowances)

	£000s
Turnover (revenue)	
Hire of equipment and services	16,500
Costs	
Materials, labour, expenses (including depreciation), overheads	14,500
Net profit	2,000
Add depreciation back in (depreciation value agreed with tax inspector)	(2,000)
Assessable profit	4,000

Deduct capital allowances on plant and equipment (agreed with tax inspector)	2,500
Add profit or (loss) from sales of fixed assets	(500)
Profit before taxation and interest	2,000
Deduct interest charges on borrowings (15% p.a.)	1,500
Profit before tax	500
Deduct corporation tax (say 35%)	175
Distributable profit after tax	325
Proposed dividend	225
Profits retained in business, C/F to Balance Sheet	100

Turnover

The company's turnover is the total value of sales of goods and services during the year, before costs are deducted. For a hire company or division this sum represents the total payments invoiced to clients for equipment hired or rented.

Assessable profit for corporation tax with capital allowances

The assessable profit is obtained by subtracting the cost of sales from the turnover value. At present published accounts are not required to reveal details of the operating costs, which comprise the following elements: materials used, that is, opening stock, plus purchases, minus closing stocks; wages, salaries and fees; expenses, excluding depreciation; administration overheads.

Depreciation and capital allowances

The principles of depreciation are discussed in Chapter 7. Depreciation amounts agreed with the Inspector of Taxes are deducted from the trading profit, hence the resulting sum gives the net profit before interest charges on borrowed capital and tax. However, internal depreciation rates are usually ignored for preparation of the Profit and Loss Account, as it is likely that adoption of capital allowances (i.e. writing down allowances), if available under tax legislation, would be more favourable. For example, taking advantage of Government policy of, say, 100 per cent capital allowances (i.e. the full cost of an item of new or used equipment set against profits during the year of purchase), would have the effect of postponing the equivalent amount of corporation tax for one year. The net effect of capital allowances thus provides opportunity to defer payment of taxes (but not avoid them, especially if there is no corresponding upsurge in profitability as a result of the purchase). Such an effect is best illustrated by a simple example of two identical companies having the same gross profit of £450, one favouring the use of capital allowances (Table 16.1), and the other not (Table 16.2). In both cases, they buy an item of plant worth £1,000, which is written off in equal amounts over four years. In both instances, the total tax paid is £280, but with Company 2 this has been deferred.

In practice it is apparent that many hire companies try to avoid paying this deferred tax by continuing to buy new equipment: year four in the second example shows what could occur if such continual buying ceased. In this case, the company would be liable to payment of taxes in excess of the net profit earned for the year. Therefore, not only do capital

Table 16.1 Company 1: not using capital allowances

	Year 1	*Year 2*	*Year 3*	*Year 4*
Profit before depreciation	450	450	450	450
Depreciation	250	250	250	250
Net profit	200	200	200	200
Tax at 35%	70	70	70	70
Profit after tax	130	130	130	130

Note
Total tax paid by Company 1 = £280.

Table 16.2 Company 2: using capital allowances

	Year 1	*Year 2*	*Year 3*	*Year 4*
Net Profit	200	200	200	200
Depreciation	250	250	250	250
Assessable profit	450	450	450	450
Capital allowance	(1,000)	(550)*	(100)*	–
Taxable profit (loss)	(550)	(100)	350	450
Tax at 35%	–	–	122.5	157.5

Notes
Total tax paid by Company 2 = £280.
* Capital allowance carried forward.

allowances provide a carrot, but also a stick. A company could, in certain circumstances, be effectively trapped in the system by a slow, almost undetected, build-up of tax liabilities.

Interest charges

Loan capital, other than shareholder's equity, usually incurs an annual interest charge. In effect, this represents an indirect cost on the business and is deducted as such from net profit for the calculation of corporation tax payment.

Profit before taxation

The profit from all the company's activities, including subsidiary and associated companies, is called the *pre-tax profit*, whereafter typically about 35 per cent of profits are subsequently payable as corporation tax. The actual tax percentage applied and due method of payment will depend upon government policy in the country of residence.

Distribution of profits after tax

After the total distributable profits are determined, amounts may be apportioned to the various claimants. The order of priorities is set out in the articles of association of the company, with loans and preference shareholders having first claim and the remainder of the distributed profits going to the ordinary shareholders. The profits awarded in this way are usually called the *dividend*. The dividend is recommended by the directors of the firm at the annual general meeting of shareholders, who then vote to accept it or otherwise. The amount of dividend will naturally depend on the level of profits and may not be declared when profits are too low. Whenever healthy profits are available for distribution, the directors usually aim for an acceptable rate of return on the par (nominal) value of the share price.

Once the dividend has been paid, the remaining profit is retained in the business and used to finance further investment in plant and other assets. Precise regulations regarding dividend distribution, however, vary depending on the country of residence and/or domicile. This can influence the amount to be declared in relation to corporation tax requirements.

The balance sheet

The function of the balance sheet is to portray the financial position of the company on a specific date – for example, 31st December of the current year. Because time is needed to prepare the Balance Sheet information, the details could change just before or immediately after publication, for example, settlement of an account by a debtor. The balance sheet is a 'photograph' of a particular financial position, whereas the profit and loss account shows the record of achievement throughout the year.

The Balance Sheet is usually presented in tabular format separating capital employed from employment of capital. Alternatively, some companies prefer liabilities and assets to be clearly separated on different pages. In either case, the two sets of measures must balance exactly. Thus, for example, any increase in cash must be counterbalanced by a decrease in some other asset or, alternatively, by an increase in liabilities. This principle applies throughout.

Example 16.2: A balance sheet with capital allowances

Using information given in the previous example together with the following company details, the Balance Sheet at 31 December may be prepared as follows:

Note 1: Information in Preparation of the Balance Sheet as at 31 December of the current year (TY)

	£000
Value of fixed assets (see Note 2)	20,000
Cash at bank	910
Stock of materials and spares	1,000
Debtors	4,000
Bank overdraft	10
Creditors	3,000
Taxation (from P. & L. Account)	175
Proposed dividend	225
Profit and loss account transferred to general reserve	100
Interest paid on loans (15% p.a.)	1,500
Issue Share Capital	1,900
Reserves at 31 December of previous year (PY)	10,000
Loan Stock (at 15% interest) at 31 December PY	12,000
Loan Stock repaid during current year	2,000
Capital allowances for plant agreed with Tax Inspector	2,500

Note 2: Valuation of Assets (Buildings and Equipment)

	£000	
Valuation at 1 January PY	21,000	
Additions in year	10,000	
Disposals during year	(3,000)	
Valuation at 31 December PY	28,000	28,000
Buildings and equipment depreciated according to company policy agreed with the Inspector of Taxes (capital allowances must be determined separately)		
Depreciation		
Cumulative up to 1 January PY	6,000	
Provided for in PY (agreed with Tax Inspector)	2,000	
	8,000	
Net book value at 31 December PY	20,000	20,000

Note 3: Capital allowances

The book value of the firm's assets shown on the Balance Sheet is normally calculated by taking a realistic rate of depreciation. Because, in this example, capital allowances and not depreciation is used to calculate profit for corporation tax assessment, the difference between capital allowances and depreciation, like unrealised profit, should be allocated to the reserves. Furthermore, some accountants would show parts of these net allowances as deferred taxation, as it should be recalled that they will be liquidated when the asset is sold, or gradually eliminated as the asset is fully depreciated over its life.

In the example given, capital allowances on equipment purchases for the current year are £2,500, whereas depreciation has been assessed at only £2,000. The net £500 has therefore been allocated to reserves.

Tabulated Balance Sheet as at 31 December of the current year (TY)

	Depreciation			
	£000	£000	£000	
Employment of capital				
Fixed assets	28,000	8,000	20,000	20,000
Current assets				
Cash at bank			910	
Stock of materials			1,000	
Debtors			4,000	
			5,910	5,910
Current liabilities				
Bank overdraft			10	
Creditors			3,000	
Taxation due			175	
Proposed dividend			225	
			3,410	3,410
Net current assets (working capital)			2,500	
			22,500	22,500

Capital employed			
Issued Share Capital Reserves		1,900	
1. P. & L. Account	100		
2. Net capital allowances	500		
3. As at 31.12.of PY	10,000		
	10,600	10,600	
Loan stock		10,000	
		22,500	22,500

Note: Because the interest paid on loans has been transacted it is not shown as a liability.

Loan capital

Short-term loans are usually classified with current liabilities, but medium- and long-term loans are placed with the capital employed. Loans may take several forms, of which debentures and loan stock form the most important sources. Such lenders of capital are creditors of the company, and not owners as are shareholders.

Long-term/medium-term finance

Long-term finance is that capital required for 5–10 years, either to start the business or to carry out expansion programmes. Broadly, the capital is used to purchase buildings, plant and equipment and to carry stocks of materials. The risks to the lender are high because of the time scale involved and, consequently, only established firms are generally considered by the lending institutions. Some of the more important sources of long-term capital are shown in Figure 16.1.

Loans are not easy to obtain. Lenders of such capital often request the borrower to provide a proportion of the finance from internal sources and, in addition, require convincing evidence that the loan capital can be secured against an asset with profitable expectations.

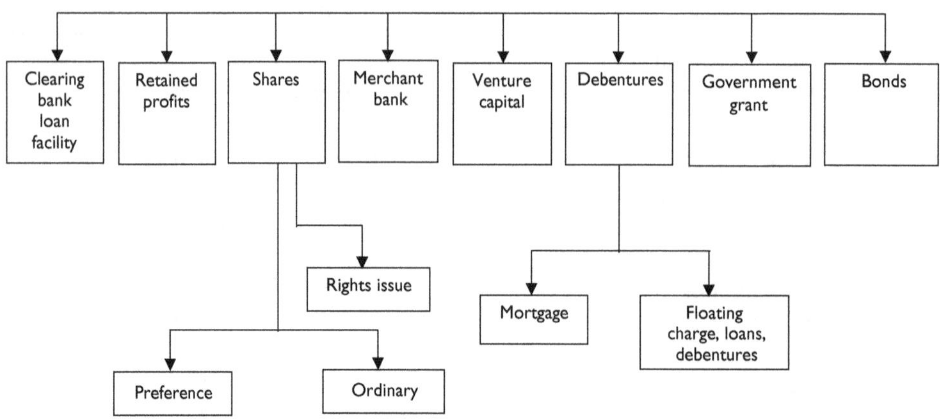

Figure 16.1 Sources of long-term finance.

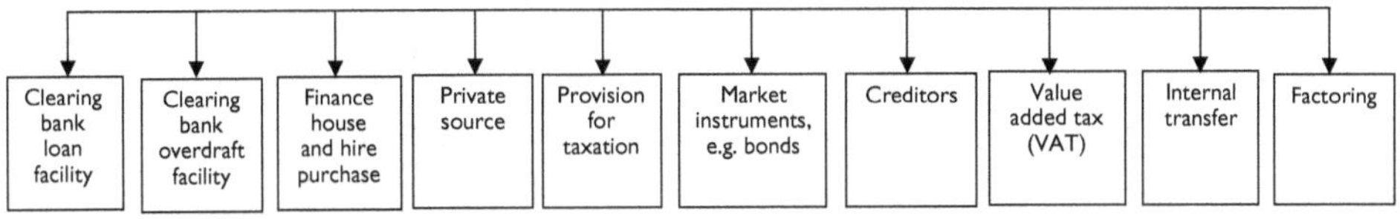

Figure 16.2 Sources of short-term finance.

The rate of interest, period of loan and capital repayments vary according to the lenders' request.

Debentures are offered for sale by the company, and are requests for loan capital with a fixed annual interest payment and life. Clearly, the company must show that it is able to pay the interest, and the capital itself will usually be mortgaged against the company's assets. Loan stock and debentures rank ahead of shareholders in entitlement for payment.

Short-term finance

The firm, when established, often needs short-term capital to overcome immediate cash flow problems. Materials have to be purchased, equipment hired, labour and subcontractors paid, and so on before payment is received for the finished product or service. Furthermore, capital may be required to smooth out the strains on cash flow resulting from rapid fluctuations in the market demand for the company's goods. Many sources of short-term finance are available to ease the situation, but naturally the firm must be well managed and profitable before the lending institutions will consider any loan application. The main sources are shown in Figure 16.2, but the clearing bank overdraft facility is the most important source. However, a leasing facility from a finance house is also an important source of either long- or short-term funding of plant acquisitions. The method is described in Chapter 6 and operates more like a rental payment than a loan.

Sources of most types of capital, their advantages and disadvantages and the costs involved are summarised in Table 16.3. Table 16.4 illustrates capital acquisition methods used by typical construction companies.

Working capital

Working capital is represented by the difference between current assets and current liabilities and is locked up in a continuous cycle, as shown in Figure 16.3.

Example 16.3: Working capital requirements

A construction company owns and operates equipment, grouped into a separate division responsible for generating its own turnover in the market place. The turnover of this division is £20 million pounds per year, broken down as 45 per cent plant ownership costs, 15 per cent for materials and 10 per cent for wages for maintenance and transport, 20 per cent for overheads incurred in maintaining the depot establishment and administration facilities; and 10 per cent profit. On average, the company keeps three months' material spares in stock and is allowed three months credit by suppliers: wages are paid weekly. Hirers of the company's equipment (i.e. debtors) are usually allowed up to two months to pay. Overheads must

Table 16.3 Summary of capital sources

Source	*Finance*	*Advantages*	*Disadvantages*	*Costs*
Bank	Overdraft	(1) Usually cheapest source (2) Quickly arranged (3) Flexible (4) No minimum (5) Renewable (6) Interest paid only on usage (7) Sometimes available unsecured	(1) Subject to changes in government economic policy (2) Repayable on demand (3) Subject to changes in bank policy (4) Tempting to use for funding long-term purchases	(1) Floating interest charge at base rate plus 1–4 per cent (2) May incur commitment fee
Bank or finance house	Short-term loan	(1) Term commitment by loan institution (2) Competition between lending houses, especially for hire purchase (3) Relatively quickly arranged (4) Can be used in conjunction with overdraft facility (5) Sometimes available unsecured	(1) Generally more expensive than overdraft (2) Term commitment and funds may therefore be idle if forecast for funds inaccurate (3) Tends to require security against other assets	(1) Floating interest charge at base rate plus 2–5 per cent
Finance house	Hire purchase facility	(1) Inexpensive and specially arranged (2) Payments fixed over term agreed (3) Ideal for short-term requirements (4) Normally overdraft facility not affected (5) Capital allowances against corporation tax available immediately (6) Not classed as borrowings	(1) Expensive (2) Subject to government economic policy changes, but never retrospectively (3) Defaults usually rigorously prosecuted (4) Interest rate quoted may be misleading, because of period of payments and compound interest calculations (5) Purchased assets not legally passed over to lender until final payment	(1) Interest fixed at time of negotiations at finance house base rate plus 4–5 per cent

Finance house	Lease facility	(1) Similar advantages to hire purchase (2) Overgeared firm can acquire resources without affecting balance sheet	(1) Ownership does not pass to lessee and therefore capital allowances not available (2) Values not reflected in balance sheet assets, and so might give distorted impression of firm's capabilities	(1) Similar to hire purchase costs, but user must be aware of possible commitments to pay for mandatory maintenance and repairs (2) Tax concessions foregone must be compared with HP alternative
Clearing banks Merchant banks	Medium term loan	(1) Term commitment by lender (2) Capital and interest repayment can be arranged to suit borrower's future cash flow position (3) Size of loan may be small or large, especially from clearing banks (4) Inflation reduces real cost over time (5) Fixed interest charge can sometimes be negotiated	(1) Usually higher interest charge than shorter-term finance (2) Long-term commitment which may require short-term borrowings to finance interest payment if cash flows became distorted from forecasts (3) Negotiation fee likely (4) Legal costs also often incurred	(1) Fixed or variable interest charge set at $1\frac{1}{2}$–4 per cent above six months inter-bank rate

Table 16.4 Capital acquisition methods used by typical construction companies

Method	Total (%)
Hire purchase	30
Lease	20
Outright purchase	50
	100

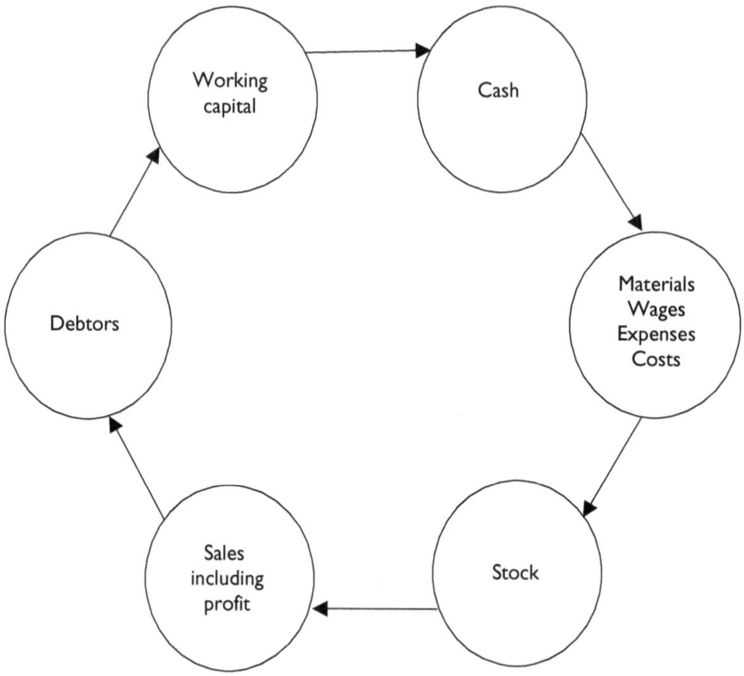

Figure 16.3 The working capital cycle.

be met monthly. The ownership costs comprise the capital repayment and interest charges for hire purchase agreements to be paid monthly. Determine the minimum working capital required.

Working capital calculations

Materials (15% × £20 millions = £3 millions p.a.)

	Time factor (in months)
Materials held in stock	3
Credit to customers	2
	5
Less credits from suppliers	3
	2

$\therefore$ Materials $= \frac{2}{12} = 3{,}000{,}000 = £500{,}000$

Labour (10% × £20 millions = £2 million p.a.)

	Time factor (in months)
Credit to customers	2
Less 1 week arrears on wages	0.25
	1.75

$$\therefore \text{ Labour} = 2{,}000{,}000 \times \frac{1.75}{12} = £292{,}000 \text{ approx.}$$

Overheads (20% × £20 millions = £4 millions p.a.)

	Time factor (in months)
Credit to customers	2
Less 1 month arrears on salaries, rents, rates, etc.	1
	1

$$\therefore \text{ Overheads} = 4{,}000{,}000 \times \frac{1}{12} = £330{,}000 \text{ approx.}$$

Ownership costs (45% × £20 millions = £9 millions p.a.)

	Time factor (in months)
Credit to customers	2
Less time delay on hire purchase payment	1
	1

$$\therefore \text{ Ownership costs} = 9{,}000{,}000 \times \frac{1}{12} = £750{,}000$$

Working capital requirements

The above figures can now be added together to calculate the total working capital requirements:

Materials	£500,000
Labour	£292,000
Overheads	£330,000
Hire purchase	£750,000
	£1,872,000

Such capital needs would usually be financed by means of an overdraft negotiated with a bank or, alternatively, with private funds. However, it can be seen that if equipment had been purchased with internal funds rather than by hire purchase, the working capital needs would be £0.75 millions less. The equivalent sum would then remain in the business and offset the working capital needs.

Aspects of the financial accounts

Ratio analysis

The main short-term techniques for managerial control are cash flow forecasting, budgetary control and costing, described in Chapters 14 and 15. Unfortunately, however, they provide absolute figures which are of marginal value in monitoring the long-term profitability and short-term liquidity of the company. Regular internal ratio analysis using the financial accounts offers a complementary post mortem to these standard procedures.

Working capital ratios

1 $\text{Current ratio} = \dfrac{\text{Current assets}}{\text{Current liabilities}}$

2 $\text{Acid test} = \dfrac{\text{Cash and debtors}}{\text{Current liabilities}}$

Most accountants look for ratios of 2:1 and 1:1, respectively. Using the Balance Sheet given earlier, it can be seen that:

$$\text{CR} = \frac{5{,}910{,}000}{3{,}410{,}000} = 1.73$$

$$\text{AT} = \frac{910{,}000 + 4{,}000{,}000}{3{,}410{,}000} = 1.44$$

In this hypothetical company the current ratio (CR) is a little low and the acid test (AT) a little high. This situation could be improved by increasing stocks by £ $\frac{1}{2}$ million and reducing creditors and debtors, respectively, by £1 million and £$1\frac{1}{2}$ million, to give the following values:

$$\text{CR} = \frac{4.91}{2.41} = 2.03$$

$$\text{AT} = \frac{3.41}{2.41} = 1.41$$

Profitability and operating ratios

1 $\dfrac{\text{Net profit before tax}}{\text{Turnover}} = \dfrac{2{,}000{,}000}{17{,}000{,}000} = 11.8\%$

2 $\dfrac{\text{Net profit before tax}}{\text{Capital employed}} = \dfrac{2{,}000{,}000}{22{,}500{,}000} = 8.9\%$ (primary ratio)

3 $\dfrac{\text{Turnover}}{\text{Capital employed}} = \dfrac{17{,}000{,}000}{22{,}500{,}000} = 0.76\%$ (turnover ratio)

The turnover ratio can provide telling information about the capital structure of the company. For example, the plant hire company is more similar to manufacturing with its turnover ratio approaching 1, whereas with most construction companies the ratio is more often nearer to 6, the difference being accounted for by the high capital investment required in plant. The ratio could, of course, be increased by leasing rather than owning equipment.

Other ratios

$$\frac{\text{Debtors}}{\text{Turnover}} \times 12 = \frac{4{,}000{,}000}{17{,}000{,}000} \times 12 = 2.28 \text{ months}$$

This means that customers are given nearly three months to pay.

$$\frac{\text{Creditors}}{\text{Purchases}} \times 12 = \frac{3{,}000{,}000}{12{,}500{,}000} \times 12 = 2.88 \text{ months}$$

This means that suppliers give nearly three months credit.

The ratios suggest that the company is applying a sensible policy with respect to both suppliers and customers, and a deeper investigation may also yield important details regarding productivity, using the following ratios:

1 $\dfrac{\text{Turnover}}{\text{Number of employees}}$

2 $\dfrac{\text{Profit}}{\text{Number of employees}}$

3 $\dfrac{\text{Plant and equipment value}}{\text{Number of employees}}$

4 $\dfrac{\text{Profit}}{\text{Plant and equipment value}}$

Capital gearing

Capital gearing is defined as the ratio of fixed return capital (FRC) to ordinary share capital, FRC being preference shares, debentures and loan stock. A company with a ratio exceeding one is described as a highly geared company. When a company can expect confidently to make a constant level of high profits over a number of years, then it is wise to raise some of the capital by means of debentures or loan stock and so improve the yield to the ordinary shareholder. Such a technique is known as leverage, and the effects are demonstrated in the following example.

Example 16.4

The capital structures for two plant hire companies are given in Table 16.5 and the profits available for interest payments and dividends for a range of company performance levels are given in Table 16.6.

1 Calculate and comment on the capital gearing ratios for each of the companies.
2 Calculate the dividend available to the ordinary shareholders and comment on the earnings per share at each level of profit. Corporation tax is 50 per cent.

Table 16.5 Capital structures

	Company X	*Company Y*
	(£1,000)	*(£1,000)*
Ordinary shares (£1)	310	70
7% Preference shares	90	90
10% Loan stock (debentures)	—	240

Table 16.6 Performance levels

	1	*2*	*3*	*4*	*5*	*6*
Profit (£)	10,000	12,000	18,000	20,000	25,000	30,000

Solution: Part (1)

	Company X	Company Y
10% debentures	–	240
7% preference shares	90	90
FRC	90	330
Ordinary shares	310	70
Capital gearing ratio	$\frac{90}{310} \times 100 = 29\%$	$\frac{330}{70} \times 100 = 470\%$
	Low-geared	High-geared

Company A is low-geared and may be able to declare a dividend to ordinary shareholders, even at low levels of profit. Company B is high-geared and must therefore ensure that profits are stable and regular and, above all, sufficient to meet the interest due on debenture holdings. However, the highly geared company should be able to yield a higher rate of return to ordinary shareholders when profits are high.

Solution: Part (2)

Table 16.7 shows that Company X is able to pay the full dividend to preference shareholders at all times: even the ordinary shareholder received payment at all the given levels of profit.

Table 16.8 shows that Company Y is in a much more difficult position while profits are poor. In fact, for the first profit period a loan or bank overdraft must be negotiated in order to meet the interest charges on the debenture stock. The full dividend to preference shareholders cannot be met until profit level five, when there is also a small amount available for distribution to ordinary shareholders. However, the high profits at levels five and six yield a far better return to the ordinary shareholders than in the low-geared company and this trend will continue for all increased levels of profit.

If the preference shares were issued as cumulative, then in the case of the highly geared situation additional dividend would be paid out in the good years to make up for the lost years, but at the expense of the ordinary shareholder.

Plant profitability

It has been argued that plant can be more advantageously organised as a profit centre and not on the service principle. This is best illustrated by the following example.

Example 14.5

A company has an annual consolidated turnover of £20 millions, of which one-fifth is from equipment rentals (refer to Table 16.9). The expected profit is 20 per cent of the capital

Table 16.7 Company X (low-geared)

	Performance level (£)					
	1	2	3	4	5	6
Ordinary shares (£1)	310	310	310	310	310	310
Preference shares	90	90	90	90	90	90
Debentures	–	–	–	–	–	–
Total capital employed (£000)	400	400	400	400	400	400
Profit before tax	10.00	12.00	18.00	20.00	25.00	30.00
Corporation tax at 35%	3.50	4.20	6.30	7.00	8.75	10.50
Profit available as dividend	6.50	7.80	11.70	13.00	16.25	19.50
Dividend on preference shares	6.30	6.30	6.30	6.30	6.30	6.30
Profit available to ordinary shareholders	0.20	1.50	5.40	6.70	9.95	13.20
Profit per ordinary share	0.06p	0.50p	1.74p	2.16p	3.20p	4.20p
Return on ordinary share capital	0.06%	0.50%	1.74%	2.16%	3.20%	4.20%
Return on total share capital	1.63%	1.95%	2.90%	3.25%	4.00%	4.80%

Table 16.8 Company Y (high-geared)

	Performance level (£)					
	1	2	3	4	5	6
Ordinary shares (£1)	70	70	70	70	70	70
7% Preference shares	90	90	90	90	90	90
5% Debentures	240	240	240	240	240	240
Total capital employed (£000)	400	400	400	400	400	400
Profit before tax	10.00	12.00	18.00	20.00	25.00	30.00
Interest on 10% debentures	12.00	12.00	12.00	12.00	12.00	12.00
Profit available	−2.00	nil	6.00	8.00	13.00	18.00
Corporation tax at 35%	nil	nil	2.10	2.80	4.55	6.30
Profit available as dividend	nil	nil	3.90	5.20	8.45	11.70
Dividend to preference shareholders	nil	nil	3.90	5.20	6.30	6.30
Profit available to ordinary shareholders	nil	nil	nil	nil	2.15	5.40
Profit per ordinary share	nil	nil	nil	nil	3.00p	7.70p
Return on ordinary share capital	nil	nil	nil	nil	3.00%	7.70%
Return on total share capital	nil	nil	2.40%	3.20%	5.28%	7.30%

Table 16.9 Plant profitability (£ millions)

	Consolidated company	*Construction business*	*Plant hire business*
Turnover	20	16	4
Capital employed	6	2	4
20% profit on capital employed	1.2	0.4	0.8
Profit expressed as a percentage on turnover	$\frac{1.2}{20} \times 100 = 6\%$	$\frac{0.4}{16} \times 100 = 2.5\%$	$\frac{0.8}{4} \times 100 = 20\%$

employed and the company is typical of the construction sector in that capital is turned over eight times per year, whereas for plant the turnover ratio is unity or thereabouts.

If plant, for instance, had made only £0.5 millions profit, then, for the company as a whole, construction would have to make £0.7 millions profit, or 4.375 per cent on turnover, to achieve 6 per cent overall on turnover. Thus, any shortfall in plant profitability requires a monumental effort from other business to redress the balance.

Accounting for inflation

Because of inflation the results portrayed by the historical method of accounting described earlier in this chapter may not be 'true and fair', as required by the Companies Act. The main effects of inflation are as follows:

1 Assets are undervalued. If, for example, a company purchases an excavator for £10,000, then, 10 years later, the same machine may cost £20,000 because of inflation. The valuation shown in the accounts, however, will be based on £10,000. The company's assets and the shareholders' interest are, therefore, undervalued. Also, more importantly, because the depreciation is based on the purchase date figure, profits will be overstated. When capital allowances (i.e. 100 per cent depreciation during the year of purchase) are in operation, this latter detrimental effect of inflation is significantly lessened.
2 Stocks of material are undervalued. The change in the value of stocks, between opening and closing dates in the accounts, may have been caused by both a volume difference and a change in prices. This latter aspect is not taken into account in calculating the cost of sales and so, in times of rising prices, the profit is overstated and extra tax is paid.
3 Companies which finance their operations by loan capital will gain over companies that are self-financing because monetary debts will depreciate in real value. For example, in real terms £1,000 borrowed and repayable 10 years later, is equivalent to £385 if the rate of inflation is 10 per cent p.a.

The Accounting Standards Committee recommends that the traditional historical accounts also be presented as a footnote with the inflation-adjusted ones: in this way the effects on profits of a combination of inflation and the present requirements for payment of corporation tax can be more clearly monitored. The necessary adjustments to the historical accounts are largely those described under points 1, 2 and 3 above, and the reader is referred to the Committee's guidance notes for further advice.

Chapter 17

Information technology in equipment management

Introduction

The management of an equipment division within a company, an equipment company within a group of companies or an independent equipment hire company encompasses all the general management tasks faced by most companies. As in other types of companies, the use of computers has become established in basic bookkeeping or accounting functions such as payroll and ledgers and in management accounting and management information systems. Examples of these uses include the maintenance of an asset register, which is particularly valuable because of the high value of capital assets represented in a large number of equipment items. Other examples include equipment location reports, cost and revenue reports, and variance analyses, which compare income with budgeted income and actual usage with planned usage. Stock control and maintenance record keeping are also examples of management information applications.

Other uses of computers in equipment management include financial appraisals where financial modelling systems are used to explore the cash flow and profitability prospects of various proposals.

IT applications in the management of equipment

Computers are used in the management of equipment in the following applications:

1. Basic accounting or bookkeeping: (a) payroll; (b) purchase ledger; (c) sales ledger; and (d) nominal ledger.
2. Management accounting and information: (a) asset register; (b) purchase and disposal analysis; (c) weekly hire charges reports; (d) period hire charges reports; (e) plant locations; (f) revenue and cost reports; (g) plant utilisation reports; (h) stock control; and (i) maintenance records.
3. Financial appraisals.

Basic accounting or bookkeeping

Payroll

Within the off-highway plant and equipment sector payroll requirements are complicated because of the diverse nature of trading enterprises (owner-operators, plant hire companies, civil engineering contractors and so forth). Very few enterprises pay all their employees on the same basis: instead there is a mixture of remuneration by piecework, and hourly, daily or weekly wages, as well as monthly salaries.

There is a great number of general-purpose payroll packages available for practically every computer in the market. As with all commercial software, payroll programs can be bought or leased for implementation on the user's own computer.

An important aspect of payroll programs is that they deal with something which is not stable. Taxation changes every year, often more than once; so does national insurance. Pension and bonus schemes also change, and so does the form and the kind of information that the Inland Revenue, and other authorities, want presented to them at the end of each year. A company's employees often work on different contracts within a single period, and the distribution of their time among these contracts changes every day. Payroll packages have to cope with this fluidity.

The basic facilities of a payroll system include:

- Payroll creation, the creation and amendment of fixed details.
- Payroll calculation, calculating the weekly and monthly pay of employees.
- Printing payslips and non-employees list.
- Amending tax and N.I. rates and pensions.
- Payment by cash, cheque or credit transfer.

Payroll systems based on timesheet information allow labour costs to be allocated to each job, thus providing information for job costing.

Sales, purchase, nominal ledger and VAT

The facilities offered in sales and purchase ledger systems are: (1) accounts creation, amendments and deletions; (2) accounts postings; (3) cash postings; (4) period statements and remittance advices. Derived from these basic accounting functions, the reports that are available include accounts over credit limit, accounts overdue, accounts turnovers and customer lists. Linked to the sales and purchase ledgers are VAT maintenance systems. These collect and analyse VAT collected and paid, enabling the total amount due to be calculated. The nominal ledger systems maintain a set of nominal accounts.

An important aspect of having the basic accounting functions computerised is that the information contained in the accounting system can be made available to other functions such as equipment and job costing and stock control. The information contained in the payroll system and in the sales and purchase ledgers is relevant to these other functions and in many accounting systems offers the important facility of feeding information to these other functions. Two important chains can be identified: purchase ledger–payroll–job costing and purchase ledger–order processing–stock control.

Management accounting and information

The applications of computers in management accounting and information are less standardised than the basic accounting functions. Consequently, there are fewer computer packages available for purchase. Also, it is a group of applications that more closely reflects the management and organisation of a company. Consequently, this group of applications, more than the basic accounting functions, have been developed within companies for their own use. Nevertheless, several software houses do offer systems purposely designed and suitable for equipment companies. The applications that come in this group are reviewed as follows.

Asset register

Assets accounting deals with assets which, usually, have a long lifespan, extending over several accounting periods. Computer programs for assets accounting provide, at any one time, (1) an assets register with straightforward information on what assets the company has, and where they are; (2) their written-down value, and depreciation to date; and (3) their replacement cost.

Information provided on the replacement cost is perhaps the most valuable, as it forms the essential test on the wisdom, or otherwise, of the depreciation policy being adopted. Most asset register systems offer a range of depreciation methods, including straight line and declining balance.

Thus, reports from asset register systems include:

- A list of plant items grouped by type of equipment.
- Inventory numbers.
- Value.
- Depreciation to date by usual methods.
- Depreciation for this year.

Separate reports will include technical details of each item of equipment held.

The importance of an asset register for the equipment company is that it maintains detailed information relating to the items of equipment held in a form that supports the production of the company accounts and in a form that allows managers to consider their depreciation and disposal policies. The depreciation policy is particularly important to an equipment company, because the depreciation element, or capital cost, forms as much as 50 per cent of a hire rate. The purchase and disposal of equipment is equally important to the continued well-being of a company.

Purchase and disposal analysis

Linked to the asset register is a purchase and disposal analysis, which is a maintained record of equipment items bought and sold. This aids the preparation of annual accounts and assists companies in managing the size of their equipment fleet.

Equipment reports

The equipment hire company requires its own set of management reports to inform the company managers and to allow them to take the various day-to-day decisions.

The management reporting systems dealing with the equipment include:

- Weekly hire charges.
- Period hire charges.
- Equipment locations.
- Revenue and cost reports.
- Equipment utilisation reports.

Here the IT systems are being employed in handling relatively large quantities of data regularly, to provide the equipment manager with the information required.

The two key reports are the Equipment Utilisation Report and the Equipment Cost and Revenue Report. These two reports provide essential information for the running of any large holding of equipment. Essentially they are designed for the equipment hire company to monitor their costs and revenues and, hence, their profits. But they are equally essential for other enterprises that hold equipment in order that they may manage the equipment subsidiary as a profit centre whose main revenues come from internal charges.

The equipment utilisation report

This information is usually produced monthly and is intended to be a statement of the equipment's actual usage in comparison with the budgeted (or planned) usage. Using the principles of management accounting, the equipment utilisation report tends to present the information in the form of variances. These variances compare the actual performance with the anticipated performance, the performance being measured in terms of the number of hours on hire and the 'price' (or revenue) obtained for these hours of hire.

Equipment cost and revenue report

This report, also produced at least monthly, is the comparison of costs and revenues. The costs listed in such a report would refer to direct costs associated with the equipment item, and so the net revenue remaining still has to pay for company overheads before being properly classified as 'profit'.

The other reports that are available include those listed here.

Weekly and period hire charges

This reporting system records the current hire charges for all equipment and the dates when hire charges were changed.

Equipment locations

Equipment location reports record the current location of equipment on hire and the length of time at that location. The location of idle equipment is also monitored in reports of this type.

Stock control

In equipment management, stock control relates mainly to the supply of spare parts and consumables. The purpose of keeping stock is to supply the fleet with the items it needs, but this need must be tempered with economy. Overstocking causes locked-up capital but understocking may cause a repair to be delayed and involve expensive downtime. Good stock control aims at achieving a balance between overstocking and understocking. Stock recording and control systems monitor stock issues and delivery notes, and perform stock taking and prompt reordering when minimum levels of stock of a particular item are reached.

Stock control systems are usually linked to purchasing systems and hold details of the various suppliers, the purchase order details for the suppliers and the invoices. Stock control systems offer savings by optimising the level of stock held and holding up-to-date information on the stocks and reducing the manpower committed to stock taking.

Maintenance records

The maintenance records can be held on computer file for ease of update and reference.

Other applications: financial appraisals

A common application relating to equipment management that is not part of day-to-day company management routines is financial appraisals. A number of investment appraisal programs are available to perform such tasks.

These investment appraisal programs provide a range of calculations which enable a proposed investment to be evaluated. The investment in this case is a capital sum from which revenue, or some other form of return, is expected, and the evaluation determines whether these returns are enough to justify the capital investment. The adequacy of these investments is measured by profitability, not just simple profit. The most common profitability measures, based on discounting techniques, are net present value and yield or rate of return. Other measures produced by these programs, which do not rely on discounting, are the payback period and the average annual rate of return.

The more sophisticated programs offer facilities which enable the modelling of the estimates and calculations that determine the investment projects' cash flow. Using this modelling facility, the estimates can be varied and the sensitivity of the investment project's profitability to certain estimates can be determined.

Communication and data exchange tools

The most significant impact that technology has had on the management of information resources within the off-highway plant and equipment sector is perhaps in the area of communications. A catalogue of communication tools that are finding increasing use within the sector, is described in the following sections.

The Internet

The Internet evolved around the latter part of the 1970s from research originally funded by the US Advanced Research Projects Agency in the late 1960s and early 1970s. Today's Internet is a global resource connecting millions of users in a labyrinth that comprises a network of computer networks. The Internet is based on a set of rules for data exchange called the TCP/IP protocol suite [which stands for Transmission Control Protocol (TCP) and Internet Protocol (IP)], an agreed method of communication between all parties associated with the Internet. In one sense, the Internet represents a community of people who use and develop the isolated computer networks, and provide a collection of resources that can be reached from those networks on a global basis. Common uses of the Internet include information sharing, interaction and communication. While the networks that make up the Internet are based on a standard set of protocols, the Internet also has gateways to networks and services that are based on other protocols. The value of the Internet to enterprises that own plant derives from its ability to connect easily and globally to a vast amount of data, which would otherwise take more time and money to organise. The following benefits can be gained by exploiting the resources of the Internet.

- Reduced communication costs.
- Enhanced co-ordination and communication.
- Acceleration in the distribution of knowledge resources within and without the company.
- Promotion and marketing for the company.

The take-up of the facilities available through the Internet by companies has however been slow. This is due to a number of technological as well as social constraints often associated with the Internet. They include security concerns, technology issues, legal uncertainties, and social acceptance by company executives and enterprise owners.

Intranets

An Intranet is a communication infrastructure that is based on the communication and content standards of the Internet but is internal to an organisation. The tools employed to create Intranets are identical to those used for the Internet and its applications. The distinguishing feature of an Intranet is that access to information is restricted to only the company's personnel. The development of Intranets was in direct response to the concerns of business regarding data security on the Internet. Intranets evolved at about the same time as the Internet revolution commenced. Companies can set up an Intranet to allow project managers access to data from both central databanks and different projects.

Extranets

An extranet is a network that uses Internet protocols and the public telecommunication system for communicating both privately and selectively with the user's customers and business partners. The technology allows a contractor to securely share part of its company information resources or operations with suppliers, subcontractors, project partners, clients or other companies. Extranets therefore, introduce an additional functionality to Intranets. The main benefit of this technology is the acceleration of business activities between different companies. Within the larger plant enterprise, the advantage in deploying extranets derives from a cheaper and more efficient way for employees and managers to connect with their suppliers, subcontractors and other project partners. For example, suppliers can receive proposals, submit bids, provide documents, and in some cases collect payments through an extranet site. Not only does this cut down on redundant ordering processes and keep suppliers up to date on future deliveries and design changes, but it also allows quicker response times to suppliers' problems.

Extranets can be used to exchange large volumes of data, including the sharing of product catalogues, providing design specifications and details, distribution of news to trading partners, and collaborating with other contractors on joint project schemes. Electronic Data Interchange (EDI) is a special form of extranet that can be set up between a contractor and suppliers.

Using the Internet

The Internet is accessed through one of the following three technologies. These are Electronic mail (E-mail), File Transfer Protocol (FTP) and the World Wide Web.

Electronic mail

Electronic mail (e-mail) is perhaps the most popular use of the Internet. The basic concepts behind e-mail are similar to that of regular mail. Documents are sent by mail to personnel in organisations at their particular addresses. In turn, they write back to a return mail address. Electronic mail however, has a distinct advantage over regular mail, which is speed. Instead of taking several days, the message can reach the recipient in minutes or perhaps even seconds (depending on location and the state of the connection to the recipient). It can be argued that the telephone achieves a similar speed of transmission. However, the e-mail provides the additional advantage of convenience and a record of communication. Messages are sent when it is convenient. Equally, the recipients respond at their convenience. Electronic mails can also be used to transmit documents as an attachment to messages. By means of e-mail, details of work sections can be exchanged between designers and contractors.

File Transfer Protocol (FTP)

One of the first information delivery systems was the File Transfer Protocol (FTP). FTP enables the electronic transfer of a whole file from one computer to another without altering the file format. This is achieved by depositing the files in a repository described as FTP site. Although e-mail systems can also transfer files, the use of FTP options presents greater flexibility. First of all, FTP avoids restrictions on the type of file sent, in particular, whether it is a binary (non-text) file, such as program software and pictures. Second, there are no restrictions on the size of file as is the case with e-mail. The transfer of engineering drawings, which are normally of large file sizes, from one section of a company to another or between project partners can readily be achieved with this technology.

World Wide Web

The World Wide Web (WWW or just 'Web' for short) is a networked hypertext information system, originally devised at CERN in Geneva, but now in use all over the world. CERN is the European Laboratory for Particle Physics founded in 1954. The basic idea of the Web is to merge the techniques of computer networking and hypertext into a powerful and easy-to-use global information system. Hypertext is text with links to further information, on the model of references in a scientific paper or cross-references in a dictionary. With electronic documents, these cross-references can be followed by a mouse-click, and within the World Wide Web environment, the references can lead to anywhere in the world. Electronic documents are similar to the pages of paper documents. A computer is used to display the pages on its screen. Inside each page sensitive spots are exploited by the computer to switch automatically from one page to another when the user clicks on a sensitive spot. This navigation by wandering from one page to another is called *browsing*. Figure 17.1 provides a typical browser interface for accessing the Web. The Web is 'seamless' in the sense that a user can see the whole Web of information as one vast hypertext document. There is no need to know where information is stored, or any details of its format or organisation. Behind this apparent simplicity there is, of course, a set of ingenious design concepts, protocols and conventions that control its manner of working.

The World Wide Web is a hypermedia information and communication system popularly used on the Internet computer network with data communications operating according to

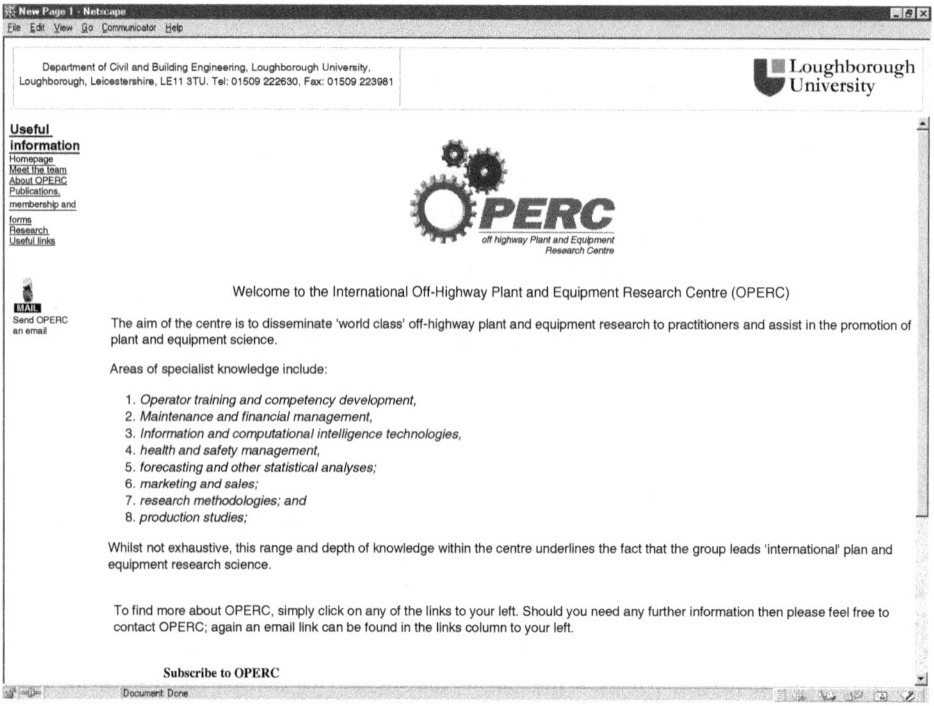

Figure 17.1 Typical browser interface for accessing the WWW.
Source: www.operc.com

a client/server model. Web clients (browsers) can access multi-protocol and hypermedia information from servers (possibly using helper applications in conjunction with the browser) by way of an addressing scheme.

Other uses of the Internet

Use of the Internet is not restricted to the transfer of text and graphic information only. It is also used for communication between people in text, audio and visual format. This is achieved through teleconferencing. Teleconferencing is technology that allows a group of people to confer simultaneously from different locations. The technology was originally based on the telephone. In recent times however, most teleconferencing is undertaken through computer-based communication such as the electronic mail. Examples include data conferencing and video conferencing.

Data conferencing

Data conference resources allow interaction between parties only to confer over text and graphic documents. This technology has already found some applications in construction, whereby designers located in different geographical regions work on the same drawing simultaneously.

Another form of data conferencing is Internet Relay Chat (IRC). IRC provides a way of communicating in *real time* with people from all over the world. It consists of various separate networks of servers that allow users to connect to a particular IRC. Generally, the user (such as you) runs a program (called a 'client') to connect to a server on one of the IRC nets. The server relays information to and from other servers on the same network. The main disadvantage associated with the IRCs is that it only employs text for communication.

Video conferencing

In its simplest form, video conferencing is the live connection of two or more people using some combination of video, audio and data for the purpose of communication, with video being the only prerequisite to fulfil the definition. Nevertheless, it is helpful to think solely of the visual impact of the medium from the outset because the roots of how we communicate with one another predate the development of language. Non-verbal actions often described as body language, essential for effective communication, are lost in the absence of visual interaction.

Rapidly developing video conferencing tools are also changing the way projects are run. Virtual conferencing tools like CU-SeeMe, Microsoft's Net Meeting and Netscape's Collabra are now being bundled with standard computer software packages, and these tools will enable project teams to collaborate, read on-line drawings and solve problems without having to travel to the job site. These advancements may help reduce travel costs and improve project communications.

Electronic commerce

Electronic commerce is employing information technology to improve transactions between companies. It involves the integration of e-mail and similar technologies into a comprehensive electronic-based system of business functions. It is based on the electronic processing and transmission of data, including text, sound and video. It encompasses many diverse activities including electronic trading of goods and services, on-line delivery of digital content, electronic fund transfers, electronic invoicing, payments and receipts, electronic share trading, collaborative design and engineering, on-line sourcing of specifications, public procurement, direct client marketing and after-sales service. It involves both products (e.g. technical goods such as material components in construction) and services (e.g. information services, financial and legal services); traditional activities (e.g. healthcare, education) and new activities (e.g. virtual malls).

Although electronic commerce is not an entirely new phenomenon, its potential impact on industry is yet to be realised and will lead to profound structural changes in industry. The take-up of electronic commerce within the off-highway plant and equipment sector of industry is being accelerated by the exponential growth of the Internet. This will grow to become transactions on a global scale between an ever-increasing number of participants, corporate and individual, known and unknown, on global open networks such as the Internet. Electronic commerce, of course, is not limited to the Internet. It includes a wide number of applications in the narrow-band (videotex), broadcast (teleshopping), and off-line environment (catalogue sales on CD-ROM), as well as proprietary corporate networks (banking). Also, the Internet is generating many innovative hybrid forms of electronic commerce by combining, for example, digital television infomercials with Internet response mechanisms (for immediate ordering), CD-ROM product catalogues with Internet connections (for content or price updates), and commercial Web sites with local CD-ROM extensions (for memory-intensive multimedia demonstrations).

Global positioning system

Throughout the 1960s, the US Navy and Air Force worked on a number of systems that would provide navigation capability for a variety of applications (e.g. tracking military vehicles and personnel). However, early systems were largely incompatible with one another and hence, in 1973, the American Department of Defense sought to unify them for the US air force. The new system developed was called Navstar Global Positioning System and was based upon atomic clocks carried on satellites. It has since come to be known simply as GPS.

Global Positioning System utilises a network of satellites placed into orbit by the United States Department of Defense. GPS was originally intended for military applications, but in the 1980s, the US government made the system available for civilian use. GPS works in any weather condition, anywhere in the world, 24 h a day. The first GPS satellite was launched in 1978 and a full constellation of 24 satellites was achieved in 1994. Each satellite is built to last about ten years and replacements are constantly being built and launched into orbit.

The nominal GPS operational constellation consists of 24 satellites that orbit the earth in 12 h. The satellite orbits repeat almost the same ground track (as the earth turns beneath them) once each day. There are six orbital planes (with nominally four satellites in each), equally spaced (60 degrees apart), and inclined at about 55 degrees with respect to the equatorial plane. This constellation provides between five and eight satellites that are visible from any point on the earth.

Global Positioning System satellites transmit signals to GPS receivers on the ground; GPS receivers passively receive satellite signals; they do not transmit. GPS receivers require an unobstructed view of the sky, so they are used only outdoors and they tend to perform poorly within forested areas or near tall buildings. GPS operations depend on a very accurate time reference, which is provided by atomic clocks at the US Naval Observatory. This is because, each GPS satellite transmits data that indicates its location and the current time. All GPS satellites synchronise operations so that these repeating signals are transmitted at the same instant. The signals, moving at the speed of light, arrive at a GPS receiver at slightly different times because the satellites are different distances from the receiver. The distance to the GPS satellites can be determined by estimating the amount of time it takes for their signals to reach the receiver.

A GPS receiver must be locked onto the signal of at least three satellites to calculate a 2D position (latitude and longitude) and track movement. With four or more satellites in view, the receiver can determine the user's 3D position (latitude, longitude and altitude). Having determined the user's position, the GPS can calculate other information, such as speed, bearing, track, trip distance, distance to destination, sunrise and sunset time etc.

The use of GPS is widespread in many fields that require geospatial information for managing assets over large areas. Forestry, mineral exploration, and wildlife habitat management all use GPS to precisely define positions of important assets and to identify changes in machine condition. For example, within agriculture GPS receivers installed onto farm equipment have provided accurate position information that enables farmers to apply fertilisers and harvest crops with great precision. GPS is however, becoming commonplace in off-highway vehicles; basic applications provide emergency on-site assistance at the push of a button (by transmitting the position of a vehicle to an emergency response centre). More sophisticated systems can pinpoint a vehicle's position on an electronic map display and create a route, giving precise directions. This has been particularly useful in tracking machines that have been stolen; a major problem for the off-highway plant and equipment industry. Other examples include fitting GPS to large dump trucks to provide accurate position information that has been used to increase productivity and enhance safety through the implementation of driverless, as well as driver operated, technology.

Appendix: Interest and time relationships and tabulations for interest rates of 10 and 15 per cent[1]

Interest and time relationships

Cash flows are transfers of money. Positive cash flows are transfers into a project or scheme, and negative cash flows are transfers of money out of a project or scheme. If a company purchased an item of equipment, the purchase price and the running costs would be negative cash flows. The revenue from hire and the resale would be positive cash flows. Cash flows can be individual lump sums or a recurring series – that is, the same cash flow recurring each period. The manipulation of both types of cash flow, the lump sum and the recurring series, is achieved by the use of derived relationships between interest and time. The derived relationships are tabulated as factors which can be used to achieve these manipulations. The expressions used to calculate these factors are given later in Tables A.1 and A.2 for the interest rates of 10 and 15 per cent. The expressions are also given in the following six examples, which are represented in the same order as that in which they are tabulated. The expressions are easily incorporated into computer programs of programmable calculations, and the use of tables is diminishing.

Compound amounts

If a sum of money, £1,000, is invested for some years – say eight – at an interest rate of 10 per cent, the sum of money that can be withdrawn at the end of that time would be:

$$£1{,}000 \times 2.1435 = £2{,}143.50$$

The amount 2.1435 is the compound amount factor, taken from the tables or calculated from the expression $(1 + i)^n$, where i is the interest rate 0.1 (for 10 per cent) and n is the number of years, 8. This process is represented in Figure A.1, to illustrate the reward received for money loaned or invested.

Present worth

The inverse of calculating compound amounts is calculating the present worth. If a sum of money, £2,143.50, is required in eight years' time the capital sum that would be required

1 10 and 15 per cent values have been given purely for illustrative purposes only. These rates will invariably change in time, and from country to country, but the method used in these calculations will remain valid.

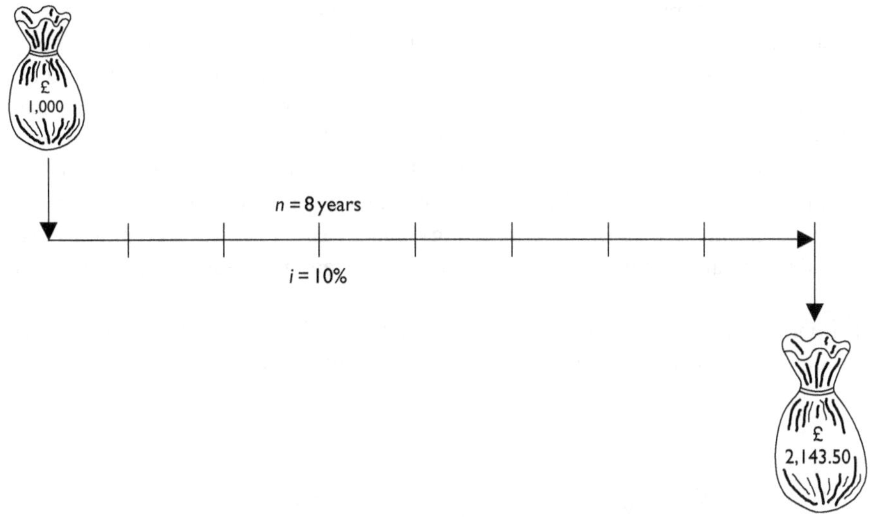

Figure A.1 Compound amount.

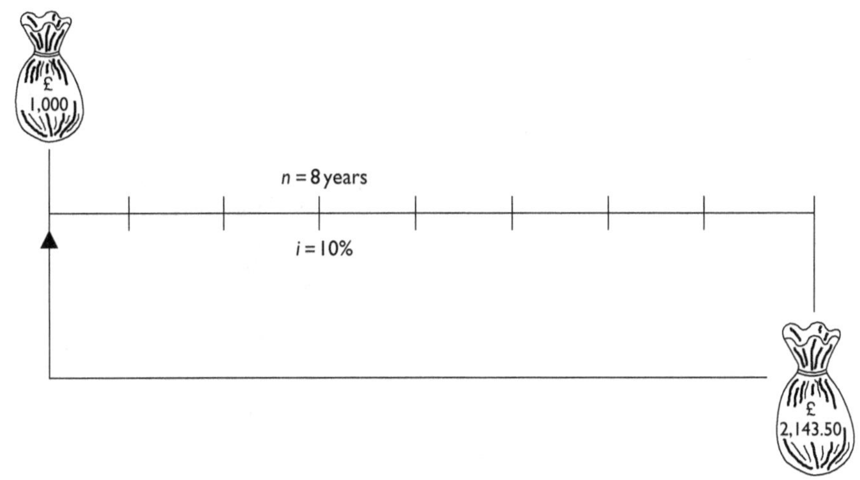

Figure A.2 Present worth.

to be invested today to generate this amount, given an interest rate of 10 per cent, would be £1,000:

$$£2,143.50 \times 0.46650 = £1,000$$

The amount 0.46650 is the present worth factor, taken from the tables or calculated from the expression:

$$\frac{1}{(1+i)^n}$$

where i is the interest rate and n is the number of years, as before. This expression is the inverse of the expression for the compound amount factor and the process can be represented as in Figure A.2, which illustrates that if £2,143.50 is required in eight years, £1,000 has to be invested now. The £1,000 is said to be the equivalent of £2,143.50 in eight years, given the interest rate of 10 per cent. Thus, £1,000 in year zero is the same as £2,143.50 in year eight and the difference of £1,143.50 is the interest earned in the intervening eight years. This process of converting the £2,143.50 in year eight to £1,000 in year zero is known as discounting. This discounting process is very widely used, because it provides a convenient method of converting future cash flows to a common base date and provides a means of comparing cash flows of different magnitude occurring at different times.

It is important to remember that the present worth of a future cash flow is the capital sum that would need to be invested today to generate that future sum.

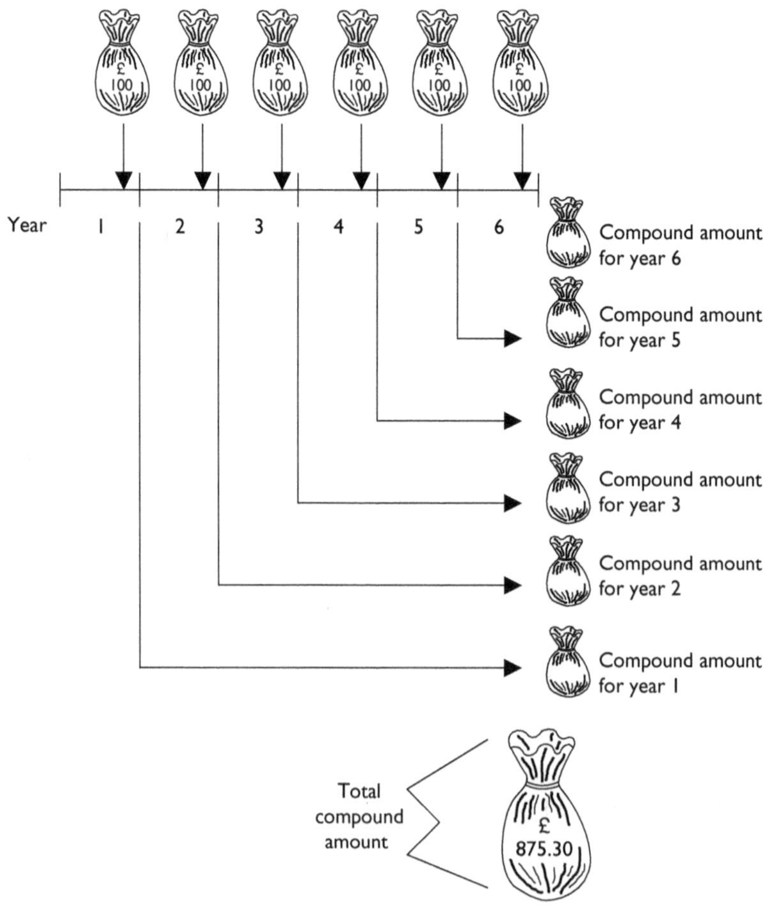

Figure A.3 Compound amount of a uniform series.

Compound amount of a regular or uniform series

If a sum of money, £100, is invested regularly for, say, six years, the sum available at the end of the six years – given an interest rate of 15 per cent – is £875.30, given by:

$$£100 \times 8.753 = £875.30$$

The amount of 8.753 is the uniform series compound amount factor, taken from the tables or calculated from the expression:

$$\frac{(1+i)^n - 1}{i}$$

where i is the interest rate and n is the number of years.

Each sum is invested for a different period of time, and the compound amount could be calculated by using the compound amount factor for each of the six individual cash flows and summing the results. The uniform series compound amount factor allows this to be done in one step, as illustrated in Figure A.3.

Figure A.3 also shows that the cash flows are included in the calculation at the end of the year or period in which they occur. This is the assumption on which the expression was derived and is known as the 'end of period convention', which states that all cash flows take place at the end of the period.

Sinking fund deposit

The inverse of the compound amount of the uniform series is to find out how much should be deposited each year or period in order to generate a certain sum at the end of the time. If £875.30 is required at the end of year six and it is intended to provide this sum by saving or investing a sum for each of the next six years at an interest rate of 15 per cent, the sum to be saved would be £100:

$$£875.30 \times 0.11423 = £100$$

The amount 0.11423 is the sinking fund deposit factor, taken from tables or calculated from the expression:

$$\frac{i}{(1+i)^n - 1}$$

where i is the interest rate and n is the number of years. This expression is the inverse of the uniform series compound amount factor. This factor is used in calculating the monies to be taken from revenue to repay borrowed capital or to replace plant.

Present worth of a regular or uniform series

As the present worth factor calculated the present worth for a lump sum, this factor calculates the present worth of a series of cash flows recurring for a number of years or periods.

The present worth of £100 each year for the next four years, given an interest rate of 15 per cent, is £285.49:

$$£100 \times 2.8549 = £285.49$$

The amount 2.8549 is the uniform series present worth factor, taken from tables or calculated from the expression:

$$\frac{(1 + i)^n - 1}{(1 + i)^n i}$$

where i is the interest rate and n is the number of years. It should be noted that the expression is the uniform series compound amount factor multiplied by the present worth factor.

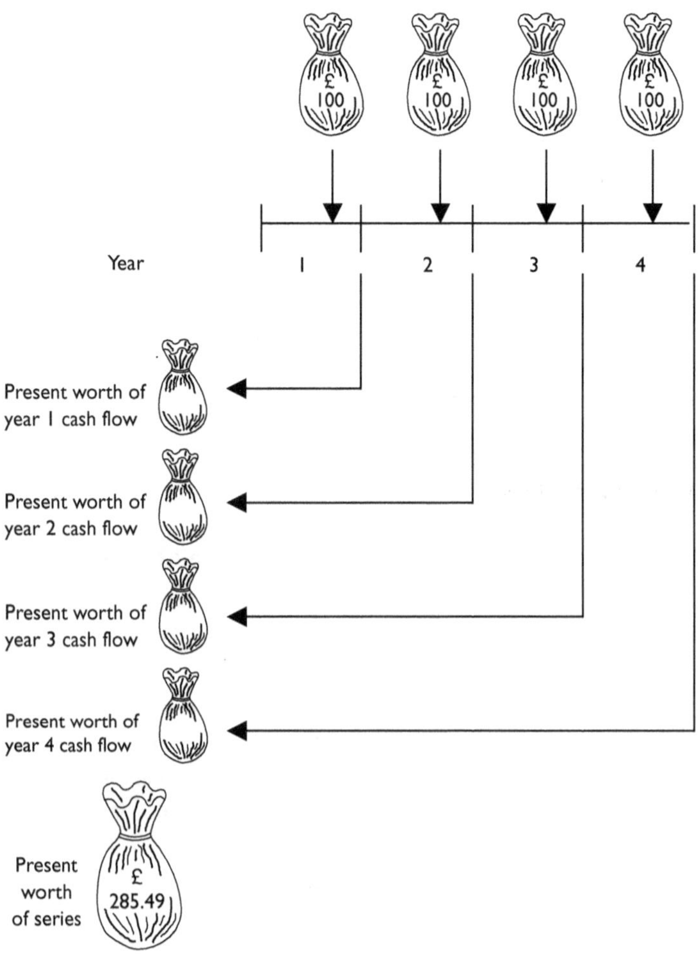

Figure A.4 Present worth of a uniform series.

The sum of £285.49 could have been found by calculating the present worth of each of the four individual cash flows and summing them. The uniform series present worth factor achieves this in one step, as illustrated in Figure A.4. The use of the end of period convention is also illustrated.

The meaning of the present worth of £285.49 for this series of cash flows is the same as that for the present worth of a single sum – that is, the sum required to be invested in year zero at 15 per cent that will allow £100 to be withdrawn each year for the next four years.

This process is widely used to discount regular series into lump sums – that is, to capitalise the regular series. It is important because it permits recurring cash flows to be converted to capital sums and thus be compared with capital sums. It provides the mechanism for comparing invested capital with running costs and thereby evaluating any possible trade-offs between them.

Capital recovery

The inverse to the uniform series present worth factor is the capital recovery factor. This allows the income that can be taken on a regular basis from an invested capital sum to be calculated. If a sum £285.49 is invested at an interest rate of 15 per cent, then the regular income that can be taken each year for the next four years is £100:

$$£285.49 \times 0.35026 = £100$$

The amount 0.35026 is the capital recovery factor taken from tables or calculated from the expression:

$$\frac{i(1+i)^n}{(1+i)^n - 1}$$

where i is the interest rate and n is the number of years. This expression is the inverse of the uniform series present worth factor.

The importance of this calculation is that it allows the conversion of capital sums into annual sums. This provides an alternative means of comparing capital with running costs. For example, if £285.49 is not invested but used to purchase an item of equipment, then the cost of that equipment item can be regarded as £285.49 or as an annual cost of £100 for four years, the annual cost being set equal to the amount of income that the investor is deprived of because the capital is locked up in the equipment. This calculation is useful in comparing hiring with purchasing or in calculating the 'capital' element in hire rates.

Tabulations of interest and time relationships for interest rates of 10 and 15 per cent

Table A.1 Interest factors for 10%

Year or period (n)	$(1+i)^n$ Compound amount of a single sum	$\frac{1}{(1+i)^n}$ Present value of a single sum	$\frac{(1+i)^n-1}{i}$ Compound amount of a uniform series	$\frac{i}{(1+i)^n-1}$ Sinking fund deposit	$\frac{(1+i)^n-1}{i(1+i)^n}$ Present worth of a uniform series	$\frac{i(1+i)^n}{(1+i)^n-1}$ Capital recovery
1	1.1000	0.90909	1.000	1.00000	0.9090	1.10000
2	1.2099	0.82644	2.099	0.47619	1.7355	0.57619
3	1.3309	0.75131	3.309	0.30211	2.4868	0.40211
4	1.4640	0.68301	4.640	0.21547	3.1698	0.31547
5	1.6105	0.62092	6.105	0.16379	3.7907	0.26379
6	1.7715	0.56447	7.715	0.12960	4.3552	0.22960
7	1.9487	0.51315	9.487	0.10540	4.8684	0.20540
8	2.1435	0.46650	11.435	0.08744	5.3349	0.18744
9	2.3579	0.42409	13.579	0.07364	5.7590	0.17364
10	2.5937	0.38554	15.937	0.06274	6.1445	0.16274
11	2.8531	0.35049	18.531	0.05396	6.4950	0.15396
12	3.1384	0.31863	21.384	0.04676	6.8136	0.14676
13	3.4522	0.28966	24.522	0.04077	7.1033	0.14077
14	3.7974	0.26333	27.974	0.03574	7.3666	0.13574
15	4.1772	0.23939	31.722	0.03147	7.6060	0.13147
16	4.5949	0.21762	35.949	0.02781	7.9237	0.12781
17	5.0544	0.19784	40.544	0.02466	8.0215	0.12466
18	5.5599	0.17985	45.599	0.02193	8.2014	0.12193
19	6.1159	0.16350	51.159	0.01954	8.3649	0.11954
20	6.7274	0.14864	57.274	0.01745	8.5135	0.11745
21	7.4002	0.13513	64.002	0.01562	8.6486	0.11562
22	8.1402	0.12284	71.402	0.01400	8.7715	0.11400
23	8.9543	0.11167	79.543	0.01257	8.8832	0.11257
24	9.8497	0.10152	88.497	0.01129	8.9847	0.11129
25	10.8347	0.09229	98.347	0.01016	9.0770	0.11016
26	11.9181	0.08390	109.181	0.00915	9.1609	0.10915
27	13.1099	0.07627	121.099	0.00825	9.2372	0.10825
28	14.4209	0.06934	134.209	0.00745	9.3065	0.10745
29	15.8630	0.06303	148.630	0.00672	9.3696	0.10672
30	17.4494	0.05730	164.494	0.00607	9.4269	0.10607
35	28.1024	0.03558	271.024	0.00368	9.6441	0.10368
40	45.2592	0.02209	442.592	0.00225	9.7790	0.10225
45	72.8904	0.01371	718.904	0.00139	9.8628	0.10139
50	117.3908	0.00851	1,163.908	0.00085	9.9148	0.10085

Note
All examples presented in this appendix are based on each period being 1 year. A period can be any other unit of time – for example, quarters, months or weeks. Care is required to ensure that the interest rate i and the unit of time for each period are compatible. For example, if the time period is months, then the interest rate must be a monthly rate.

Table A.2 Interest factors for 15%

Year or period (*n*)	$(1+i)^n$ *Compound amount of a single sum*	$\frac{1}{(1+i)^n}$ *Present value of a single sum*	$\frac{(1+i)^n-1}{i}$ *Compound amount of a uniform series*	$\frac{i}{(1+i)^n-1}$ *Sinking fund deposit*	$\frac{(1+i)^n-1}{i(1+i)^n}$ *Present worth of a uniform series*	$\frac{i(1+i)^n}{(1+i)^n-1}$ *Capital recovery*
1	1.1500	0.86956	1.000	1.00000	0.8695	1.15000
2	1.3224	0.75614	2.149	0.46511	1.6257	0.61511
3	1.5208	0.65751	3.472	0.28797	2.2832	0.43797
4	1.7490	0.57175	4.993	0.20026	2.8549	0.35026
5	2.0113	0.49717	6.742	0.14831	3.3521	0.29831
6	2.3130	0.43232	8.753	0.11423	3.7844	0.26423
7	2.6600	0.37593	11.066	0.09036	4.1604	0.24036
8	3.0590	0.32690	13.725	0.07285	4.4873	0.22285
9	3.5178	0.28426	16.785	0.05957	4.7715	0.20957
10	4.0455	0.24718	20.303	0.04925	5.0187	0.19925
11	4.6523	0.21494	24.349	0.04106	5.2337	0.19106
12	5.3502	0.18690	29.001	0.03448	5.4206	0.18448
13	6.1527	0.16252	34.351	0.02911	5.5831	0.17911
14	7.0757	0.14132	40.504	0.02468	5.7244	0.17468
15	8.1370	0.12289	47.580	0.02101	5.8473	0.17101
16	9.3576	0.10686	55.717	0.01794	5.9542	0.16794
17	10.7612	0.09292	65.075	0.01536	6.0471	0.16536
18	12.3754	0.08080	75.836	0.01318	6.1279	0.16318
19	14.2317	0.07026	88.211	0.01133	6.1982	0.16133
20	16.3665	0.06110	102.443	0.00976	6.2593	0.15976
21	18.8215	0.05313	118.810	0.00841	6.3124	0.15841
22	21.6447	0.04620	137.631	0.00726	6.3586	0.15726
23	24.8914	0.04017	159.276	0.00627	6.3988	0.15627
24	28.6251	0.03493	184.167	0.00542	6.4337	0.15542
25	32.9189	0.03037	212.793	0.00469	6.4641	0.15469
26	37.8567	0.02641	245.711	0.00406	6.4905	0.15406
27	43.5353	0.02296	283.568	0.00352	6.5135	0.15352
28	50.0656	0.01997	327.104	0.00305	6.5335	0.15305
29	57.5754	0.01736	377.169	0.00265	6.5508	0.15265
30	66.2117	0.01510	434.745	0.00230	6.5659	0.15230
35	133.1755	0.00750	881.170	0.00113	6.6166	0.15113
40	267.8635	0.00373	1,779.090	0.00056	6.6417	0.15056
45	538.7692	0.00185	3,585.128	0.00027	6.6542	0.15027
50	1,083.6573	0.00092	7,217.715	0.00013	6.6605	0.15013

Bibliography

Barlow, J.F. (1999) Excel Models for Business and Operations Management, Wiley, London.

Boddy, D. and Paton, R. (1998) Management: An Introduction, Prentice Hall, New Jersey.

Brimson, T. (1995) The Health and Safety Survival Guide: A Comprehensive Handbook of Managers, McGraw-Hill, London.

BS 1139 Part 1 1981, Part 3 (1983) Specification for Prefabricated Access and Working Towers, British Standards Institute, London.

BS 5744 (1974) Code of Practice for the Safe Use of Cranes, British Standards Institute, London.

BS 5974 (1990) Code of Practice for Temporary Installed Suspended Scaffolds and Access Equipment, British Standards Institute, London.

BS 6031 (1981) Code of Practice for Earthworks, British Standards Institute, London.

BS 6187 (1982) Code of Practice for Demolition, British Standards Institute, London.

BS 7171 (1989) Specification for Mobile Elevating Work Platforms, British Standards Institute, London.

BS 7212 (1989) Code of Practice for the Safe Use of Construction Hoists, British Standards Institute, London.

Chatfield, C. (1996) The Analysis of Time Series: An Introduction, 5th ed., Chapman and Hall, London.

Clarke, T. (1999) Managing Health and Safety in Building and Construction, Butterworth Heinemann, London.

Day, D.A. and Benjamin, N.B.H. (1991) Construction Equipment Guide, 2nd ed., Wiley, New York.

Dixon, J.R. (1966) Design Engineering – Inventiveness Analysis and Decision Making, McGraw-Hill, London.

Edwards, D.J. (2002) A Voluntary Code of Practice for Plant and Equipment Operator Training and Competence Development, Loughborough University Press, Distributed by IRAS Group (sales@irasgroup.net).

Edwards, D.J. and Griffiths, I.J. (2000) An Artificial Intelligent Approach to Calculating Hydraulic Excavator Cycle Time and Productivity Output, Transactions of the Institute of Mining and Metallurgy, Vol. 109, pp. A23–29.

Edwards, D.J. and Holt, G.D. (2000) ESTIVATE: A Model for Calculating Hydraulic Excavator Productivity and Output Costs, Engineering, Construction and Architectural Management, Vol. 7, No.1, pp. 52–62.

Edwards, D.J., Holt, G.D. and Harris, F.C. (1998) Maintenance Management of Heavy Duty Construction Plant and Equipment, Chandos Publishing, Oxford.

Grimshaw, P. (1985) Excavators, Blandford Press, Dorset.

Harris, F.C. (1994) Modern Construction and Ground Engineering Equipment, Longman, London.

Haddock, K. (1998) Giant Earthmovers, An Illustrated History, MBI Publishing, Osceola, WI, USA.

Illingworth, J.R. (1998) Construction Methods and Planning, 2nd ed., E & FN Spon, London.

Johnson, B. (1999) Classic Plant and Machinery, MacMillan, London.

Kvanli, A.H., Guynes, C.S. and Pavur, R.J. (2000) Introduction to Business Statistics: A Computer Integrated Approach, 4th ed., West Publishing, New York.

Lapin, L.L. (1994) Quantitative Methods for Business Decisions with Cases, 6th ed., Duxbury Press, London.

Leffingwell, R. (1994) Caterpillar Farm Crawlers and Bulldozers, MBI Publishing, Osceola, WI, USA.

Lester, A. (2000) Project Planning and Control, 3rd ed., Butterworth Heinemann, London.

Levine, D.M., Krehbiel, T.C. and Berenson, M.L. (2000) Busines Statistics: A First Course, 2nd ed., Prentice Hall, New Jersey.

Mitchell, J. (1998) JCB the First 25 Years (1945–1995), 2nd ed., Special Event Books, Salisbury.

Nunnally, S.W. (2000) Managing Construction Equipment, 2nd ed., Prentice Hall, New Jersey.

Ober, G.J. (1999) Operating Techniques for the Tractor Loader Backhoe, Equipment Training Resources, Northridge, CA, USA.

Orlemann, E.C. Super-duty Earthmovers, MBI Publishing, Osceola, WI, USA.

Parr, A. (1998) Hydraulics and Pneumatics: A Technician's and Engineer's Guide, 2nd ed., Butterworth Heinemann, London.

Peurifoy, R., Ledbetter, W. and Schnexnayder, C. (1996) Construction Planning, Equipment and Methods, McGraw Hill, 5th ed., New York/London.

Pilcher, R. (1992) Principles of Construction Management, 3rd ed., McGraw-Hill, London.

Russell, J.E. (1985) Construction Equipment, Prentice Hall, New Jersey.

Index

www.ingramcontent.com/pod-product-compliance
Lightning Source LLC
LaVergne TN
LVHW080642011026
849160LV00014B/1293
* 9 7 8 0 4 1 5 2 5 1 2 8 0 *